THE APPLICATIONS OF GENERALIZED FUNCTIONS

A collection of papers presented at the Symposium on "The Applications of Generalized Functions" sponsored by the Air Force Office of Scientific Research at the 1966 Fall Meeting of Society for Industrial and Applied Mathematics held at the State University of New York at Stony Brook

September 12-14, 1966

Society for Industrial and Applied Mathematics
Philadelphia, Pennsylvania, 1967

Reprinted from
SIAM Journal on Applied Mathematics,
July 1967, pages 771–1111.

CONTENTS

PREFACE

It has been some seventeen years now since L. Schwartz's books, "Théorie des Distributions," appeared. His work has had a profound impact on mathematical analysis, especially on the theory of partial differential equations. Indeed, there has been a proliferation of research into the various theories of generalized functions and their implications in functional analysis, but most of this research has been concerned with the "purer" aspects of mathematics. Nevertheless, generalized functions have also been of importance in a number of physical and engineering disciplines, most notably in quantum field theory and passive system theory.

In accordance with this, the theme of the 1966 Fall meeting of the Society for Industrial and Applied Mathematics, held at the State University of New York at Stony Brook, was chosen to be "The Applications of Generalized Functions." One purpose of the meeting was to provide a forum for the discussion of research trends and open problems in this subject. The papers contained in this issue of the *SIAM Journal on Applied Mathematics* is a result.

The first three papers herein, namely, those by Bremermann, de Jager and Güttinger, were invited lectures at the meeting and survey various aspects of the theme. The remaining papers report some recent research results. Those by Anderson and Newcomb and by Dolezal are concerned with problems arising from electrical network theory, whereas Beltrami's work is perhaps better classified in the more general field of mathematical system theory. The paper by Güttinger and Pfaffelhuber is concerned with quantum field theory, whereas the note by Jaffe was motivated by his research in this field. Liverman's work relates the concept of a distribution to the measurement of a physical variable. Finally, the papers of Beltrami and Wohlers and of Zemanian are concerned with generalized integral transforms and their uses, a subject of some importance in applied mathematics.

These papers are merely a sampling of the various applications of generalized functions. Most surely, many pertinent aspects of the subject do not appear here. It is hoped, however, that the publication of these papers in this issue of the *SIAM Journal on Applied Mathematics* will help to stimulate further research into the uses of generalized functions in the physical sciences, thereby making them still more useful as a tool of analysis for scientists and engineers.

A. H. Zemanian

SOME REMARKS ON ANALYTIC REPRESENTATIONS AND PRODUCTS OF DISTRIBUTIONS*

H. J. BREMERMANN†

1. Representation of distributions by analytic functions of one complex variable. This section is a brief summary; most of the proofs may be found in Bremermann [5]. Compare also Beltrami-Wohlers [1]. Terminology and notations are consistent with L. Schwartz [12].

Let T be a distribution in $(\mathcal{E}'(E^1))$, where E^1 denotes the reals, or $T \in (\mathcal{D}'_{L_2})$; then the function $\hat{T}(z) = (1/2\pi i)\langle T, 1/t - z\rangle$ is an analytic function for $\operatorname{Im} z \neq 0$ ($\operatorname{Im} z$ = imaginary part of z). $\hat{T}(z)$ represents T in the following sense: $\hat{T}(x + i\epsilon) - \hat{T}(x - i\epsilon)$ converges to T for $\epsilon \to 0+$ in the topology of $(\mathcal{D}')$. We call $\hat{T}(z)$ the *Cauchy representation* of T.

For $T \in (\mathcal{D}'(E^1))$ the function $(t - z)^{-1}$ is not a test function, $\hat{T}(z)$, in general, does not exist, but the following result holds: there exists a pair of functions $f_+(z)$ and $f_-(z)$, analytic in the upper and lower half-plane, respectively, such that $f_+(x + i\epsilon) - f_-(x - i\epsilon)$ converges to T for $\epsilon \to 0+$ in the topology of $(\mathcal{D}')$. We call f_+ the *forward function*, f_- the *backward function* and the pair an *analytic representation* of T. Analytic representations of the same distribution differ by at most an entire function [14], [24].

If a pair of analytic functions represents T, then their complex derivatives represent T'.

If the complement of the support of a distribution is nonempty, then for any analytic representation of T the forward and backward functions f_+ and f_- are analytic continuations of each other and remain analytic on the complement of the support of T.

Not every pair of functions, analytic in the upper and lower half-planes, respectively, represents a distribution: $f_+(z) = f_-(z) = \exp(1/z^2)$ is a counter-example.

A *tempered function* is a continuous complex-valued function on E^1 that grows at infinity no faster than a polynomial. For a tempered function f the integrals $\int_0^\infty f(t)e^{izt}\,dt$ and $-\int_{-\infty}^0 f(t)e^{izt}\,dt$ converge for $\operatorname{Im} z > 0$ and $\operatorname{Im} z < 0$, respectively. We call this pair of functions the *Carleman-Fourier transform*. (In Bremermann [5] it is called "generalized Fourier transform". The new notation is less pallid and gives tribute to Carleman's pioneering

* Received by the editors November 4, 1966. Presented by invitation at the Symposium on "The Applications of Generalized Functions" sponsored by the Air Force Office of Scientific Research at the 1966 Fall Meeting of Society for Industrial and Applied Mathematics held at the State University of New York at Stony Brook, September 12–14, 1966.

† Department of Mathematics, University of California, Berkeley, California 94720.

work [8].) It is an analytic representation of the Schwartz-Fourier transforms of f. If $f \in L_2$, then it coincides with the Cauchy analytic representation of the Plancherel-Fourier transform of f. Any tempered distribution $(T \in (\mathcal{S}'))$ is a finite order derivative of a tempered function. The Carleman-Fourier transform of $f^{(m)}$, where f is a tempered function, is defined as $(-iz)^m$ times the transform of f, and this is an analytic representation of the Schwartz-Fourier transform of $f^{(m)}$.

The Carleman-Fourier transform of $T \in (\mathcal{S}')$ behaves at infinity at most like a polynomial and at finite points $x_0 \in E^1$ at most like $(z - x_0)^{-m}$ for some finite m. Conversely, a pair of functions with these properties is an analytic representation of a tempered distribution (Vladimirov [15]).

If an analytic representation vanishes in the lower half-plane and satisfies the asymptotic conditions, then it is the Carleman-Fourier transform of a distribution with support in the nonnegative real axis. Conversely if a distribution in $(\mathcal{S}')$ has its support in the nonnegative real axis (we denote this space by $(\mathcal{S}_+'))$, then its Carleman-Fourier transform vanishes in the lower half-plane.

The Carleman-Fourier transform of a tempered function with nonnegative support is identical to the Laplace transform if we substitute $-s$ for iz.

Operations with the Carleman-Fourier transforms are sometimes simpler than operations with the conventional single-sided Laplace transform. (Compare Roberts-Kaufmann [11].) For example, when transforming a differential equation with constant coefficients, $L(f', s) = sL(f, s) + f(0)$ (for the Laplace transform L), while $\mathfrak{F}(f', z) = (-iz)\mathfrak{F}(f, z)$ (for the Carleman-Fourier transform $\mathfrak{F}$). Thus, in particular, when higher order derivatives occur, the Carleman-Fourier transform leads to simpler expressions.

It would be formally advantageous to solve electrical network and control system problems by means of Carleman-Fourier techniques instead of Laplace transforms, except for the weight of tradition. (See Bremermann [5, Chaps. 10 and 11]. Compare also Beltrami-Wohlers [1] whose distributional Laplace transform is identical with our Carleman-Fourier transform except for a factor i in the complex variable.)

2. Products of Fourier transforms of distributions in $(\mathcal{S}_+')$. *Multipliers* are classes of functions for which a product with all distributions from a given space may be defined (cf. Bremermann [5]). All (C^∞) functions are multipliers for $(\mathcal{D}')$, and conversely, if a function is a multiplier for all distributions in $(\mathcal{D}')$, then it is a (C^∞) function.

For individual distributions, products with factors other than multipliers may sometimes be defined. In particular, the Carleman-Fourier transforms of distributions in $(\mathcal{S}_+')$ form an algebra, which we denote by $\hat{K}^+$. Indeed, the product of two analytic functions in $\hat{K}^+$ belongs again to $\hat{K}^+$ and thus corresponds to a distribution in $(\mathcal{S}_+')$. We denote the algebra of the corresponding distributions by K^+, consistent with the notation of Vladimirov [15]. The product in K^+ corresponds to convolution in $(\mathcal{S}_+')$: $\mathfrak{F}(S, z)\mathfrak{F}(T, z) = \mathfrak{F}(S*T, z)$ for S, $T \in (\mathcal{S}_+')$.

If the (ordinary) boundary values of two functions in $\hat{K}^+$ exist, then their product is an analytic representation of the ordinary product of the boundary value functions. Thus the distribution product defined in this way is consistent with ordinary multiplication.

The distributions in K^+ have the property that they cannot vanish in an open set without being identically zero. If $T \in K^+$, then there exists a representation f_+, f_- such that $f_- \equiv 0$. If the complement of the support of T were nonempty, then f_+ would tend to zero on a nonempty open set of the real axis, which would imply $f_+(z) \equiv 0$, which would imply $T \equiv 0$. Thus, in particular, all distributions with compact support (such as Dirac's δ and its derivatives) are excluded from the algebra K^+.

For distributions T within K^+, on the other hand, not only powers but certain analytic functions can be defined. If T is represented by $f_+ \in \hat{K}^+$, then we define $g(T)$ by means of $g(f_+(z))$. The requirement on g is simply that $g(f_+(z)) \in \hat{K}^+$. For example, the distribution δ_+ is defined by $\delta_+ = (1/2\pi)\mathfrak{F}(H)$, where H is the Heaviside step function, $\mathfrak{F}$ the Schwartz-Fourier transform, $\delta_+ \in K^+$, and where

$$\frac{1}{2\pi}\mathfrak{F}(H, z) = \begin{cases} -\dfrac{1}{2\pi i z} & \text{for } \operatorname{Im} z > 0, \\ 0 & \text{for } \operatorname{Im} z < 0. \end{cases}$$

Thus we can define $\sqrt{\delta_+}$, $\log \delta_+$, etc., as indicated. Unfortunately, one cannot define $\sqrt{\delta}$, $\log \delta$, etc., in this way.

3. A multiplication for tempered distributions. As L. Schwartz has shown, there is no multiplication for $P(1/x)$ (principal part of $1/x$), x, and δ that is associative and that is consistent with multiplier multiplication. x is a multiplier. $P(1/x)x = 1$, $x\delta = 0$; 1 and 0 are multipliers; thus $[P(1/x)x]\delta = \delta$ but $P(1/x)[x\delta] = 0$.

The distributions $P(1/x)$, x, and δ are all in $(\mathcal{S}')$. Giving up uniqueness of the product, we can define an associative and commutative multiplication for $(\mathcal{S}')$ as follows.

Let S, $T \in (\mathcal{S}')$. Let f_+, f_- be a representation of S, let g_+, g_- be a representation of T, with f_+, g_+ in $\hat{K}^+$. (f_+, f_-) is unique up to an entire function (§1). The condition $f_+ \in \hat{K}^+$ implies *uniqueness up to a polynomial.* Let $S \cdot T$ be the distribution that is defined by (f_+g_+, f_-g_-). It is clear that f_+g_+ defines a distribution in K^+ and f_-g_- in K^- (which is defined analogously to K^+). Thus $S \cdot T$ is defined and in $(\mathcal{S}')$. Since (f_+, f_-) and (g_+, g_-) are unique up to an additive polynomial, $S \cdot T$ is defined up to $S \cdot p + T \cdot C$, where p and q are arbitrary polynomials. For example, if $S = \delta$, $T = \delta'$, we have $S \cdot T$ defined up to $\delta(a_0 + \cdots + a_n x^n) = a_0 \delta$ and $\delta'(b_0 + \cdots + b_m x^m) = b_0 \delta' - b_1 \delta$. Observing that

$$\hat{\delta}(z) = \frac{-1}{2\pi i z}, \qquad \hat{\delta}'(z) = \frac{1}{2\pi i z^2},$$

we obtain

$$\delta \cdot \delta' = \frac{1}{4\pi i} \delta'' + C_0 \, \delta + C_1 \, \delta',$$

where C_0 and C_1 are arbitrary constants.

In §6 we describe a rather different method of obtaining products (by means of formally differentiating Fourier transforms). It leads to the same arbitrary part $C_0 \delta + C_1 \delta'$, but gives zero instead of $\delta''/4\pi i$.

Unfortunately, the multiplication just described is not consistent with ordinary multiplication for functions. One might be tempted to define *multiplication by means of* $(f_+ - f_-)(g_+ - g_-)$, but unfortunately $f_-(x - i\epsilon)g_+(x + i\epsilon) + f_+(x + i\epsilon)g_-(x - i\epsilon)$, in general, does not converge to a distribution. One could argue that one should simply drop the bothersome term (as in other regularization procedures) and define the product by $f_+(x + i\epsilon)g_+(x + i\epsilon) + f_-(x - i\epsilon)g_-(x - i\epsilon)$ (instead of having a minus sign between the two terms). However, under the first definition the product of distributions *with compact support* is a distribution with compact support, while under the second definition this is not the case. For example, for δ we have

$$f_+ (z) = f_- (z) = \frac{1}{2\pi i z}.$$

$$f_+^2(x + i\epsilon) - f_-^2(x + i\epsilon) \to \frac{1}{2\pi i} \delta',$$

while

$$f_+^2(x + i\epsilon) + f_-^2(x - i\epsilon) \to \frac{-1}{2\pi} P\left(\frac{1}{x^2}\right),$$

where $P(1/x^2)$ is the principal value of $1/x^2$. The support of $P(1/x^2)$ is the entire real axis. The undetermined part is in both cases the same, namely, $c\delta$, where c is an arbitrary constant.

For distributions with compact support we may use the Cauchy representation to define the product. The Cauchy representation is uniquely determined by this requirement: $f_+(z), f_-(z) \to 0$ for $z \to \infty$. In this way we get a unique product but no consistency with ordinary multiplication and hence no consistency with multiplier multiplication. This multiplication has also been studied by Itano [21]. As in the case of K^+ we can define functions of distributions for suitable analytic functions, which, however, seems less meaningful than for K^+.

4. Analytic representation in several variables. Let $T \in (\mathcal{E}'(E^n))$; then

$$\hat{T}(z) = \left\langle T, \left[(2\pi i)^n \prod_{j=i}^{n} (t_j - z_j) \right]^{-1} \right\rangle$$

is an analytic function of n complex variables for $\operatorname{Im} z_j \neq 0, j = 1, \cdots, n$, $z = (z_1, \cdots, z_n)$. The sum of 2^n terms

$$\hat{T}(x_1 + i\epsilon, \cdots, x_n + i\epsilon) - \hat{T}(x_1 - i\epsilon, \cdots, x_n + i\epsilon) + \cdots (-1)^n \hat{T}(x_1 - i\epsilon, \cdots, x_n - i\epsilon)$$

converges to T for $\epsilon \to 0+$ in the topology of $(\mathcal{D}')$ (see Bremermann [5]). We call $\hat{T}(z)$ the Cauchy representation of T.

Other than in the case of one variable the singularities of $\hat{T}(z)$ are not contained in the support of T, and they depend upon the choice of the coordinates $x_1, \cdots, x_n$ in E^n. $\hat{\delta}(z) = \dfrac{1}{[(2\pi i)^2 z_1 z_2]}$ is an example.

We define *tempered functions* analogously to $n = 1$, and the *Carleman-Fourier transform* (called generalized Fourier transform in [5]) of a tempered function f is defined by

$$\mathfrak{F}(f, z) = \begin{cases} \displaystyle\int_0^\infty \int_0^\infty \cdots \int_0^\infty f(t) e^{it\cdot z}\, dt & \text{for } \operatorname{Im} z_1 > 0, \cdots, \operatorname{Im} z_n > 0, \\ \displaystyle -\int_{-\infty}^0 \int_0^\infty \cdots \int_0^\infty f(t) e^{it\cdot z}\, dt & \text{for } \operatorname{Im} z_1 < 0, \operatorname{Im} z_2 > 0, \cdots, \operatorname{Im} z_n > 0, \\ \displaystyle (-1)^n \int_{-\infty}^0 \int_{-\infty}^0 \cdots \int_{-\infty}^0 f(t) e^{it\cdot z}\, dt & \text{for } \operatorname{Im} z_1 < 0, \cdots, \operatorname{Im} z_n < 0. \end{cases}$$

Here $t \cdot z$ denotes $t_1 z_1 + \cdots + t_n z_n$. The Carleman-Fourier transform of

$$\frac{\partial^{p_1 + \cdots + p_n}}{\partial t_1^{p_2} \cdots \partial t_n^{p_n}} f,$$

f tempered, is defined by

$$(-iz_1)^{p_1} \cdots (-iz_n)^{p_n} \mathfrak{F}(f, z).$$

It is an analytic representation of the Schwartz-Fourier transform of f, and for L_2-functions it coincides with the Cauchy representation of the Plancherel-Fourier transform.

5. Fourier transforms of distributions vanishing outside the lightcone. The Cauchy representation involves 2^n terms, a large number for all but small n, and depends upon the choice of coordinates $x_1, \cdots, x_n$. The representation by Wightman functions, described in the following, involves only two terms. It is invariant under Lorentz transforms if the distribution represented is Lorentz-invariant. On the other hand, it is applicable only to distributions that are Fourier transforms of distributions vanishing outside the lightcone.

Let:

$$C^+ = \{x \mid x_0^2 - x_1^2 \cdots x_n^2 > 0, x_0 > 0\},$$

$$C^- = \{x \mid x_0^2 - x_1^2 \cdots x_n^2 > 0, x_0 < 0\}.$$

C^+ is called the *forward lightcone*, C^- the *backward lightcone*. $C^+ \cup C^-$ is called the *lightcone*. Let

$$T^+ = \{z \mid y \in C^+, x \text{ arbitrary}\},$$

$x = (x_0, \cdots, x_n)$, $y = (y_0, \cdots, y_n)$, $z = x + iy$. T^+ is called the *forward tube domain*, and $T^- = \{z \mid y \in C^-, x \text{ arbitrary}\}$ the *backward tube domain*.

A tempered distribution with support in $\overline{C^+}$ has a (Schwartz) Fourier transform that is the boundary limit of a function analytic in T^+, and if the support is in $\overline{C^-}$, the Fourier transform is the boundary limit of a function analytic in T^-. A Fourier transform of a distribution with support in the lightcone is thus represented by a pair of analytic functions. We call such analytic representations *Wightman functions.*

If $F(t)$ is a tempered function supported in the lightcone, then

$$\int_{C^+} F(t)e^{it\cdot z}\, dt, \qquad -\int_{C^+} F(t)e^{it\cdot z}\, dt$$

is a pair of Wightman functions for the (Schwartz) Fourier transform of F. ($t\cdot z$ stands for $t_0z_0 + t_1z_1 + \cdots + t_nz_n$.)

The most general tempered distribution is a finite order derivative D^p of a tempered function F; a pair of Wightman functions is obtained by multiplying $\int_{C^+} F(t)e^{it\cdot z}\, dt$ and $-\int_{C^-} F(t)e^{it\cdot z}\, dt$ by $(-iz)^p = (-iz_1)^{p_1} \cdots (-iz_n)^{p_n}$.

If $F(t)$ is an L_2-function, then its Plancherel-Fourier transform $g(\xi)$ has a representation given by Bochner's integral:

$$\frac{i^{(n+1)}\Gamma((n+1)/2)}{2\pi^{(n+3)/2}} \int_{E^{n+1}} \frac{g(\xi)}{[(z-\xi)^2]^{(n+1)/2}}\, d\xi,$$

where

$$(z - \xi)^2 = (z_0 - \xi_0)^2 - (z_1 - \xi_1)^2 - \cdots - (z_n - \xi_n)^2.$$

The kernel, in this integral, is identical to the transform $\int_{C^+} e^{it\cdot z}\, dt$ for $z \in T^+$ and $\int_{C^-} e^{it\cdot z}\, dt$ for $z \in T^-$ (compare Bochner [2]).

In the general case of a tempered distribution S with support in the lightcone we have the following Vladimirov representation for the Schwartz-Fourier transform $\mathfrak{F}(S)$.

There exists an s_0 *such that, for all* $s \geqq s_0$, $\langle \mathfrak{F}(S), \Box_\xi^s k(\xi - z)\rangle$ *exists, is analytic for* $z \in T^+ \cup T^-$ *and represents* $\Box^s\mathfrak{F}(S)$. *Here* $\Box_\xi = (\partial^2/\partial\xi_0^2) - (\partial^2/\partial\xi_1^2) - \cdots - (\partial/\partial\xi_n^2)$, *and* $\Box^s$ *is* $\Box$ *iterated s times* (Vladimirov [15], Bremermann [5]).

If the complement of the support of $\mathfrak{F}(S)$ is not empty, then any two Wightman functions are analytic continuations of each other, thanks to the *edge of the wedge theorem*. This can be stated as follows.

THEOREM 1. *Let* f_+ *be analytic in* T^+, f_- *in* T^-. *For* $x \in D \subset E^{n+1}$ *let the boundary limits of* f_+ *and* f_- *be identical for* $y \to 0$, $|(y_1, \cdots, y_n)| \leqq \lambda |y_0|$, $0 \leqq \lambda < 1$. *Then* f_+ *and* f_- *are analytic continuations of each other and the resulting function is analytic in a neighborhood* (*with respect to* C^{n+1}) $N(D)$ *of* D (F. Browder [7], Tornehave [23], Bremermann, Oehme, Taylor [6]).

The resulting domain $T^+ \cup T^- \cup N(D)$, in general, is not a domain of analyticity. The intersection of the envelope of analyticity of $T^+ \cup T^- \cup N(D)$ intersects the real space E^{n+1}, in general, in a domain $E(D)$ larger than D.

The envelope $E(D)$ can be computed by means of the following Kontinuitätssatz.

THEOREM 2. *Let* $x^{(1)}$ *and* $x^{(2)}$ *be two points in D. If* $x^{(1)}$ *and* $x^{(2)}$ *can be connected by a curve* $x(t)$ *in D such that the line segment* $x^{(1)}x(t)$ *remains in the cone* $C^+ \cup C^- \cup \{0\}$ *as* $x(t)$ *runs from* $x^{(1)}$ *to* $x^{(2)}$ (*inclusively*), *then the line segment* $x^{(1)}$ *belongs to* $E(D)$.

The process of adjoining points can be iterated. Thus we obtain an envelope that is not convex in the ordinary sense, but, in the terminology of Vladimorov [15], [22], *convex with respect to time-wise admissible curves.*

These results can be generalized to arbitrary convex cones containing the x_0-axis. The corresponding tube domains are then to be taken over the *dual cones*, and the line segment $x^{(1)}x(t)$ of time-wise admissible curves is subject to the condition that it must stay in the *dual cone.*

For each $\mathfrak{F}(S)$, $S \in (\mathcal{S}'(\bar{C}))$, there exist Wightman functions that behave at infinity like a polynomial and such that $[(z - x_0)^2]^m f(z)$ tends to zero for $z \to x_0$ with $y \to 0$ in the lightcone, for m sufficiently large.

Distributions with $f_- \equiv 0$ form an algebra which we denote again by

K^+ and what has been said in §2 applies almost verbatim. One can also define distribution products, analogous to §3. The resulting algebra has been investigated by Vladimirov [15], [16].

The theory can be generalized from cones to symmetric domains. Bochner [2] has given his integral representations for these as well.

6. Distribution products that occur in quantum field theory. Products of distributions that have been defined so far are either for K^+; K^-, or for $(\mathcal{S}')$.

Distribution products that occur in quantum field theory, unfortunately, do not belong to K^+, K^-, but are, on the other hand, required only for a rather special subclass of $(\mathcal{S}')$: Fourier transforms of certain rational functions.

For L_2 functions we have the identity

$$f*g = \mathfrak{F}^{-1}(\mathfrak{F}(f)\cdot\mathfrak{F}(g)),$$

where $f*g$ denotes convolution, $\mathfrak{F}$ the Plancherel—Fourier transform, and $\mathfrak{F}^{-1}$ the inverse Placherel—Fourier transform. Here $\mathfrak{F}(f), \mathfrak{F}(g) \in L_2$ and hence $\mathfrak{F}(f)\cdot\mathfrak{F}(g) \in L_1$. In the case where $\mathfrak{F}(f)$ and $\mathfrak{F}(g)$ are distributions one might try to define their product by means of $\mathfrak{F}(f*g)$. This, however, only puts the shoe on the other foot. In general, it is as difficult to define $f*g$ as to define $\mathfrak{F}(f)\cdot\mathfrak{F}(g)$. In fact, in [5] *convolution* of distributions is defined *by means of* $\mathfrak{F}(f)\cdot\mathfrak{F}(g)$, under the conditions that either one of the factors is a multiplier for the other or that both $\mathfrak{F}(f)$, $\mathfrak{F}(g)$ belong to K^+ or K^-.

The Feynman integrals that occur in the S-matrix perturbation expansion of quantum electrodynamics, in general, diverge. However, divergence is only "of at most second order", thanks to the power counting theorem, Dyson [18], Bogoliubov and Shirkov [3]. If we differentiate under the integrals with respect to the free variables (external line momenta), then the second and third order Feynman integrals become convergent.

This, in the case of a convolution (and similarly for other Feynman integrals), can be interpreted in various ways. The convolution integral is defined on the subspace of $(\mathcal{S})$ consisting of all $\phi \in (\mathcal{S})$ that are derivatives of second order of functions in $(\mathcal{S})$. This method is also equivalent to defining

$$D^p\mathfrak{F}^{-1}(\mathfrak{F}(f)\mathfrak{F}(g)) = \mathfrak{F}^{-1}((iz)^p\mathfrak{F}(f)\mathfrak{F}(g)),$$

which is equivalent to defining the product $(iz)^p\mathfrak{F}(f)\mathfrak{F}(g)$ instead of $\mathfrak{F}(f)\mathfrak{F}(g)$, which is equivalent to defining $\mathfrak{F}(f)\mathfrak{F}(g)$ on the subspace of all functions $\phi \in (\mathcal{S})$ that vanish like x^p on

$$\{x_1 = 0\} \cup \{x_2 = 0\} \cup \cdots \cup \{x_n = 0\}.$$

Example. $\delta\cdot\delta' = \mathfrak{F}^{-1}(1)\mathfrak{F}^{-1}(it)$. $\mathfrak{F}(\delta\cdot\delta')$ corresponds formally to the diverging convolution $\int_{-\infty}^{\infty} 1\cdot(x - t)\,dt$. Differentiating formally twice with respect to x we obtain 0. Formally, we also have $[\mathfrak{F}(\delta\cdot\delta')]'' = -\mathfrak{F}(\delta\cdot[\delta'\cdot t^2])$. Now, $\delta\cdot[\delta'\cdot t^2] = [t^2\cdot\delta]\cdot\delta' = 0$. Hence we interpret $\mathfrak{F}(\delta\cdot\delta')$ as the solution of $[\mathfrak{F}(\delta\cdot\delta')]'' = 0$, so that $\mathfrak{F}(\delta\cdot\delta') = a + b\omega$, a and b arbitrary constants. Hence $\delta\cdot\delta' = c_0\delta + c_1\delta'$, where c_0 and c_1 are arbitrary constants. The arbitrary part agrees with that of the product defined in §3; the definite parts, however, differ.

The method just described leads us to distribution products defined on a subspace of $(\mathcal{S})$ which then may be extended to all of $(\mathcal{S})$. This extension involves some arbitrary constants; their number depends upon the dimension of the remaining space. This latter concept of distribution products defined on a subspace is a method used by König [10] in a rather general theory. The method of differentiation on the other hand gives us the subspace in a concrete, practical way, namely, as $\{\psi \mid \psi = x^p\phi,\ \phi \in (\mathcal{S})\}$. At the same time we have an alternate interpretation of the constants as constants of integration. De Jager [9] has published an investigation of different methods of "regularizing divergent convolution integrals" and he shows that the different methods of regularization suggested by Feynman, Akhiezer, and Pauli-Villars, this author, and others can all be interpreted in the same way: defining the Feynman integrals on subspaces of derivatives of $(\mathcal{S})$. Extension to the full space then involves arbitrary constants.

This author has suggested in [4] to interpret the arbitrary constants in the sense that the theoretical machinery of quantum electrodynamics is not a complete description of nature and that the values of their constants have to be supplied from experiment.

Bremermann [4] discusses explicitly only Feynman integrals of order two and three.

Dyson [18], Bogoliubov and Parasiuk [19], and Hepp [17] have shown that the Feynman integrals can be regularized consistently in all orders by means of "counter terms." Hepp has described these methods as "a constructive form of the Hahn-Banach theorem", the distribution products being defined on a subspace of $(\mathcal{S})$ and then extended. Due to the combinatorial complexity of the Feynman graphs of higher order the subspaces are not easy to keep track of. It seems to this author that the same results could be obtained perhaps in a simpler way through differentiation with respect to the momenta of external lines. (For convolutions this method was explained above.) For Feynman integrals of order two and three this method has been illustrated by Bremermann [5]. A complete proof would require a demonstration that the Feynman integrals of *all orders* can be made convergent in this way.

Still another way of regularizing Feynman integrals has been suggested by Güttinger [20], [20a].

While the physical interpretation of divergent Feynman integrals is still open to debate, we can, in many other instances (for example, the energy of a δ-pulse in network theory) ascribe the divergent quantity to an *overidealized mathematical description* of a more complicated physical reality (compare Bremermann [5]).

7. Distribution products and the entropy integral of information theory. We will now describe another problem where apparently ill-defined products and functions of distributions occur and suggest a resolution of the difficulties through a restatement of the problem and a reinterpretation of the quantities involved.

Consider a random variable x, taking real values in $(-\infty, \infty)$ with a probability distribution $\Phi(x)$, with density $\Phi'(x) = \phi(x)$. The entropy of $\phi(x)$, according to Shannon [13], is defined as

$$-\int_{-\infty}^{\infty} \phi(x) \log \phi(x)\, dx,$$

and the integral may or may not exist, depending upon $\phi(x)$.

Consider, for example, a random variable that takes the values 1 and 2 with probabilities p_1 and p_2, all other values with probability zero. Then $p_1 + p_2 = 1$, $\phi(x) = p_1\delta(x-1) + p_2\delta(x-2)$. For the entropy we have formally

$$-\int_{-\infty}^{\infty} [p_1\, \delta(x-1) + p_2\, \delta(x-2)] \log [p_1\, \delta(x-1) + p_2\, \delta(x-2)]\, dx.$$

Here the expression $\log [p_1\delta(x-1) + p_2\delta(x-2)]$ is not well defined; $p_1\delta(x-1) + p_2\delta(x-2)$ does not belong to K^+, the algebra for which we defined log in §2.

There does not seem to be a reasonable way to define $\log \delta(x)$ individually: $\delta \equiv 0$ for $x \neq 0$, thus $\log \delta$ should be $\equiv -\infty$ for $x \neq 0$. $\delta \cdot \log \delta$, however, is another matter. We can try a method that has also been used in the regularization of Feynman integrals. Approximate $\phi(x)$ by a regularized function ϕ_n and go to the limit. In our case, ϕ_n can be interpreted as the probability density of a random variable that takes not only the values 1 and 2 but other values as well, though with a small probability. For example, we can approximate the δ functions at 1 and 2 by Gaussians:

$$\delta_n(x-1) = \sqrt{\frac{n}{\pi}}\, e^{-n(x-1)^2};$$

$\delta_n(x - 2)$ is approximated analogously. Note that

$$\lim\, [\delta_n(x - 1) \log (p_1\delta_n(x - 1) + p_2\delta_n(x - 2)) - \delta_n(x - 1) \log p_1\delta_n(x - 1)] = 0.$$

Denote by H_n the regularized entropy integral

$$- \int_{-\infty}^{\infty} [p_1\delta_n(x - 1) + p_2\delta_n(x - 2)] \log (p_1\delta_n(x - 1) + p_2\delta_n(x - 1))\, dx.$$

Then,

$$\lim_{n\to\infty} \left[H_n + p_1 \log p_1 + p_2 \log p_2 + p_1 \int_{-\infty}^{\infty} \delta_n(x - 1) \log \delta_n(x - 1)\, dx + p_2 \int_{-\infty}^{\infty} \delta_n(x - 2) \log \delta_n(x - 2)\, dx \right] = 0.$$

By change of variable and because of $p_1 + p_2 = 1$ we have

$$p_1 \int_{-\infty}^{\infty} \delta_n(x - 1) \log \delta_n(x - 1)\, dx + p_2 \int_{-\infty}^{\infty} \delta_n(x - 2) \log \delta_n(x - 2)\, dx = \int_{-\infty}^{\infty} \delta_n(x) \log \delta_n(x)\, dt.$$

Thus,

$$\lim_{n\to\infty} \left[H_n + p_1 \log p_1 + p_2 \log p_2 + \int_{-\infty}^{\infty} \delta_n(x) \log \delta_n(x)\, dx \right] = 0.$$

The integral $\int_{-\infty}^{\infty} \delta_n(x) \log \delta_n(x)\, dx$ diverges. It may be tempting simply to discard this part and to define $-p_1 \log p_1 - p_2 \log p_2$ as the "finite part" of the entropy integral, in particular since $-\sum_{i=1}^{2} p_i \log p_i$ is the entropy for a discrete probability distribution. Indeed, if two "events" occur with probabilities p_1 and p_2, we can without loss of generality describe this by a real random variable taking the values 1 and 2 with probabilities p_1 and p_2. (Instead of 1 and 2 we may choose any other pair of distinct real numbers. The change of variable and $p_1 + p_2 = 1$ reduce the divergent part in any case to $\int_{-\infty}^{\infty} \delta_n(x) \log \delta_n(x)\, dx$.)

Discarding the divergent part would be quite analogous to discarding the divergent part of a regularized Feynman integral and keeping a suitably chosen finite part that gives agreement with experiments. This is done for the Feynman and Pauli-Villars regularizations (compare Bogoliubov and Shirkov [3] and de Jager [9]).

In our case, however, the divergent part can be given an interpretation. For finite n, $\delta_n(x)$ is the probability density of a random variable with mean zero and standard deviation $(2\sqrt{n})^{-1}$. (We recall that "mean zero" was obtained by a change of variable.) The entropy integral

$$-\int_{-\infty}^{\infty} \delta_n(x) \log \delta_n(x)\, dx$$

can be interpreted as the amount of information (number of bits) necessary to specify a real number to within an accuracy of (about) $(2\sqrt{n})^{-1}$. As n tends to infinity the uncertainty goes to zero, but the amount of information required tends to infinity. $\phi(x) = \delta(x - x_0)$ corresponds to the case where the random variable takes with certainty the value x_0. Writing down a real number x_0 with absolute precision takes indeed infinitely many bits. The divergent entropy integral thus, in a sense, is real, and should not be discarded.

Examining the case of finite probability again, we note that our previous interpretation was incomplete. We do not observe a random variable taking precise values but observe it with a priori knowledge that it takes one of finitely many values. We can describe this a priori knowledge as a probability distribution. The same applies to the continuous case. A random variable is observed. Before the observation a probability distribution is given (constituting the a priori knowledge, e.g., about the range of the variable). After measurement (or observation) we have a new probability distribution. Measurement error is reflected in the resulting distribution.

Thus, instead of being concerned with the information (entropy) associated with a probability distribution we should be concerned with the information *gained*, given an a priori probability distribution and an equivocation.

Following similar considerations by Shannon [13] we define the information gain (also called "mutual information") as

$$G = H(x) - H_y(x),$$

where $H(x)$ is the entropy of the a priori distribution of the random variable x, y is the observed value, and $H_y(x)$ is the equivocation

$$H_y(x) = -\int_{-\infty}^{\infty} p(y) \int_{-\infty}^{\infty} p_y(x) \log p_y(x)\, dx dy,$$

where $p_y(x)$ is the conditional probability density that x has occurred when y has been observed.

Noting that $p(y)p_y(x) = p(x, y)$, where $p(x, y)$ is the probability density of the joint event (x, y), we can write

$$H_y(x) = -\int_{-\infty}^{\infty}\int_{-\infty}^{\infty} p(x, y) \log \frac{p(x, y)}{p(y)}\, dxdy.$$

Noting that $\iint p(x, y) \log q(x)\, dxdy = \int q(x) \log q(x)\, dx$, where $q(x)$ is the probability density of x, we can write:

$$G = \int_{-\infty}^{\infty}\int_{-\infty}^{\infty} p(x, y)[\log p(x, y) - \log q(x)p(y)]\, dxdy.$$

Note that $G = H(x) + H(x, y)$, where $H(x, y)$ is the entropy of the joint event.

If the distributions of x and y are independent, then $H(x, y) = H(x) + H(y)$ and $G = 0$. Indeed, if y is independent of x, the measurement tells nothing about x. Note that G is independent of scale changes (linear transformations of x), while $H(x)$ is not.

It is possible that G exists while the individual entropies $H(x)$ and $H_y(x)$ are infinite. Consider the case of $p(x) = p_1\delta(x - 1) + p_2\delta(x - 2)$, which was discussed before.

$$\bar{H}(x) = -\sum_{i=1}^{2} p_i \log p_i$$

and

$$\bar{H}_y(x) = -\sum_{i=1}^{2} p_i \sum_{i=1}^{2} p_i(j) \log p_i(j) = \bar{H}(x, y) - \bar{H}(y)$$

denote entropy and equivocation as defined for finite probability. Here p_i is the probability of y taking the value i, q_i the probability that x takes the value i, and $p_i(j)$ the conditional probability that x takes the value j when y is i. p_{ij} is the probability of the joint event $x = i$ and $y = j$.

Regularizing we obtain for G:

$$G = -\int \sum\sum p_{ij}\delta_n(x - i)\delta_n(y - j)\, [\log \sum\sum p_{ij}\delta_n(x - i)\delta_n(x - j)$$
$$- \log \sum q_i\delta_n(x - i) - \log \sum p_j\delta_n(y - j)]\, dxdy.$$

Arguing as before we reduce the integral to

$$-\iint \sum\sum p_{ij}\delta_n(x - i)\delta_n(y - i)\, [\log p_{ij}\delta_n(x - i)\delta_n(y - j)$$
$$- \log q_i\delta_n(x - i) - \log p_j\delta_n(y - j)]\, dxdy.$$

This integral decomposes into

$$-\bar{H}(x,y) + \bar{H}(x) + \bar{H}(y) = \bar{H}(x) - \bar{H}_y(x)$$

and

$$\iint \sum\sum p_{ij}\delta_n(x-i)\delta_n(x-j)\,[\log \delta_n(x-i)\delta_n(y-j) - \log \delta_n(x-i) - \log \delta_n(y-j)]\,dxdy = 0.$$

Thus the divergent parts cancel out. Note that the same calculation applies to the case of any finite number of m events with probabilities $p_1, \cdots, p_m$, and to higher dimensions.

If we were to apply a similar reasoning to Feynman integrals we might try to interpret regularization as the subtraction of a priori information.

REFERENCES

[1] E. J. BELTRAMI AND M. R. WOHLERS, *Distributions and the Boundary Values of Analytic Functions*, Academic Press, New York and London, 1966.

[2] S. BOCHNER, *Group invariance of Cauchy's formula in several variables*, Ann. of Math., 45 (1944), pp. 686–707.

[3] N. N. BOGOLIUBOV AND D. V. SHIRKOV, *Introduction to the Theory of Quantized Fields*, Interscience, New York, 1959.

[4] J. H. BREMERMANN, *On finite renormalization constants and the multiplication of causal functions in perturbation theory*, ONR Report, Department of Mathematics, University of California, Berkeley, 1959.

[5] ———, *Distributions, Complex Variables, and Fourier Transforms*, Addison-Wesley, Reading, Massachusetts, 1965.

[6] H. J. BREMERMANN, R. OEHME AND J. G. TAYLOR, *Proof of dispersion relations in quantized field theory*, Phys. Rev., 109 (1958), pp. 2178–2190.

[7] F. BROWDER, *On the edge of the wedge theorem*, Canad. J. Math., 15 (1963), pp. 125–131.

[8] T. CARLEMAN, *L'intégrale de Fourier et questions qui s'y rattachent*, Almqvist and Wiksell, Uppsala, 1944.

[9] E. M. DE JAGER, *Divergent convolution integrals in electrodynamics*, ONR Tech. Rep., Department of Mathematics, University of California, Berkeley, 1963.

[10] H. KÖNIG, *Multiplikationstheorie der verallgemeinerten Distributionen*, Bayer. Akad. Wiss. Math.-Nat. Kl. Abh. (N. F.), no. 82, 1957, 80 pp.

[11] G. E. ROBERTS AND H. KAUFMAN, *Table of Laplace Transforms*, Saunders, Philadelphia, 1966.

[12] L. SCHWARTZ, *Théorie des Distributions*, vols. I, II, Hermann, Paris, 1957, 1959.

[13] C. E. SHANNON, *A mathematical theory of communication*, Bell System Tech. J., 27 (1948), pp. 379–423, 623–656.

[14] H. STAPP, Notes, Theoretical Group, Lawrence Radiation Laboratory, Berkeley, California, 1966. (Also: University of California Radiation Laboratory Rep. 16816, Appendix C.)

[15] V. S. VLADIMIROV, *Construction of envelopes of holomorphy for regions of a special type and their application*, Trudy Mat. Inst. Steklov., 60 (1961), pp. 101–144.

[16] ———, *Methods of the Theory of Functions of Several Complex Variables*, M.I.T. Press, Cambridge, Massachusetts, 1966.

[17] K. HEPP, *Proof of the Bogoliubov-Parasiuk theorem on renormalization*, Communicat. Math. Physics, 2 (1966), pp. 301–326.

[18] F. J. DYSON, *S-matrix in quantum electrodynamics*, Phys. Rev., 75 (1949), pp. 1736–1755.

[19] N. N. BOGOLIUBOV AND O. S. PARASIUK, *Über die Multiplikation der Kausalfunktionen in der Quantentheorie der Felder*, Acta Math., 97 (1957), pp. 227–266.

[20] W. GÜTTINGER, *Generalized Functions and Dispersion Relations in Physics*, Fortschr. Physik, to appear.

[20a] ——— , *Generalized functions in elementary particle physics and passive system theory: Recent trends and problems*, this Journal, 15 (1967), pp. 964–1000.

[21] MITSUYUKI ITANO, *On the multiplicative products of distributions*, J. Sci. Hiroshima Univ. Ser. A-I Math., 29 (1965), pp. 51–74.

[22] V. S. VLADIMIROV, *Construction of envelopes of holomorphy for a special kind of region*, Dokl. Akad. Nauk SSSR, 134 (1960), pp. 251–254.

[23] H. TORNEHAVE, *On analytic functions of several variables. Analytic continuation by Schwartz's reflexion method*, Mat. Tidsskr. B, (1952), pp. 29–37.

[24] S. RAJNAK, Private communication.

THE LORENTZ-INVARIANT SOLUTIONS OF THE KLEIN-GORDON EQUATION*

E. M. DE JAGER†

1. Introduction. The solutions, invariant under a proper Lorentz transformation, of the Klein-Gordon equation (see, e.g., [1], [2] and [3])

$$(\square - m^2)f(x) = -\delta(x) \tag{1.1}$$

play an important role in relativistic quantum field theory. The function $f(x)$ is the wave function connected with a particle of mass m, x denotes the coordinates of a point (x_1, x_2, x_3, x_0) in R_4; (x_1, x_2, x_3) are space coordinates and x_0 is the time coordinate. The symbol $\square$ stands for the differential operator

$$\frac{\partial^2}{\partial x_1^2} + \frac{\partial^2}{\partial x_2^2} + \frac{\partial^2}{\partial x_3^2} - \frac{\partial^2}{\partial x_0^2}, \tag{1.2}$$

and $\delta(x)$ denotes the four-dimensional Dirac function concentrated in the origin of the coordinate system.

In the case of $m = 0$, (1.1) reduces to the ordinary wave equation in three-dimensional space and the function $f(x)$ is the wave function connected with a photon.

In textbooks on field theory the solutions of (1.1) are usually obtained in a rather formal way. They are determined by applying a Fourier transformation to (1.1), and the transform $\hat{f}(k)$ of $f(x)$ satisfies the equation

$$(k^2 - m^2)\hat{f}(k) = -1, \tag{1.3}$$

with $k^2 = k_0^2 - k_1^2 - k_2^2 - k_3^2$. The general Lorentz-invariant solution of (1.3) is of the form

$$\hat{f}(k) = \frac{1}{m^2 - k^2} + c_+ \delta_+(m^2 - k^2) + c_- \delta_-(m^2 - k^2), \tag{1.4}$$

where $\delta_+(m^2 - k^2)$ and $\delta_-(m^2 - k^2)$ are Dirac functions concentrated on, respectively, the upper and lower sheets of the hyperboloid $m^2 - k^2 = 0$ and c_+ and c_- are arbitrary constants.

The inverse transforms of $\delta_+(m^2 - k^2)$ and $\delta_-(m^2 - k^2)$ are obtained by purely formal calculations; for example, divergent integrals are converted into convergent integrals by merely interchanging the operations of differ-

* Received by the editors November 4, 1966. Presented by invitation at the Symposium on "The Applications of Generalized Functions" sponsored by the Air Force Office of Scientific Research at the 1966 Fall Meeting of Society for Industrial and Applied Mathematics held at the State University of New York at Stony Brook, September 12–14, 1966.

† Technological University Twente, Enschede, The Netherlands.

entiation and integration (see [1, §15.1], [2, §15.b]). It is obvious that this rather formal procedure cannot claim sufficiently mathematical rigor. The difficulties stem essentially from the fact that the Dirac functions $\delta_{\pm}(m^2 - k^2)$ are not functions in the classical sense; they are generalized functions or distributions and they should be treated as such. To obtain the Lorentz-invariant solutions of (1.1) in a rigorous way one needs essentially the theory of distributions and the calculations have to be performed within the frame work of this theory (see, e.g., [4]–[8]).

In this paper we present a rigorous derivation of the Lorentz-invariant solutions of the Klein-Gordon equation; we have chosen this problem as the subject of our lecture at this SIAM conference, devoted to "applications of generalized functions to system theory", because many interesting theorems of distribution theory have to be applied in order to solve (1.1) in a rigorous way.

A proper derivation of the solutions of (1.1) has been given earlier by several authors, namely, P. D. Methée [9]–[10], J. Lavoine [12] and the present author [8, Chap. IV], [13].

Methée [9] applies a mapping of the n-dimensional space R_n on the line R, given by the transformation

$$u = x_0^2 - \sum_{i=1}^{n-1} x_i^2. \tag{1.5}$$

Lorentz-invariant solutions of the Klein-Gordon equation in n dimensions are derived by means of pairs of distributions defined on C^∞ functions with compact support in R, i.e., on the space of test functions, which is usually denoted by $\mathfrak{D}$. Asymptotic expansions of distributions, concentrated on the hyperboloid $u = \epsilon$, in the neighborhood of $\epsilon = 0$ play an important role in the theory.

A simplified version of this theory is due to J. E. Roos and L. Gårding [12]. These authors use the following transformations for the test functions $\phi(x)$:

$$(M\phi)(\tau) = \int \phi(x)\delta(\tau - x^2)\, dx, \tag{1.6}$$

$$(M_1\phi)(\tau) = \int \phi(x)\delta(\tau - x^2) \operatorname{sgn} x_0\, dx, \tag{1.7}$$

with $x^2 = x_0^2 - x_1^2 - x_2^2 - x_3^2$.

Linear homeomorphisms are established between the spaces of even and uneven Lorentz-invariant distributions and the duals of the spaces of the functions $(M\phi)(\tau)$, respectively, $(M_1\phi)(\tau)$; hence the necessary calculations can be performed in these dual spaces.

Lavoine [12] and the present author [8], [13] solve (1.1) by applying a Fourier transformation; however, the method of Lavoine for obtaining the inverse transforms of the terms appearing in (1.4) is essentially different from the one given in this paper. Whereas Methée uses distributions defined on the space $\mathfrak{D}$ of test functions with compact support, Lavoine and the present author consider only tempered distributions. This results in the fact that the solutions obtained in this paper and also in that by Lavoine, contrary to those in [9], do not have terms which increase exponentially at infinity. However, due to their behavior at infinity these terms are usually disregarded by the physicists. We have confined our treatment to four dimensions, but the theory can be extended to the general case of n dimensions.

Finally it may be remarked that this paper differs in a few details from [8, Chap. IV] and [13], since some improvements could be made.

2. Outline of the method. A proper Lorentz transformation is a linear transformation of the space R_4, which leaves invariant the quadratic form

$$x^2 = x_0^2 - x_1^2 - x_2^2 - x_3^2, \tag{2.1}$$

and does not interchange the forward and backward lightcone; moreover, its determinant equals $+1$. In the sequel we shall always write Lorentz transformation instead of proper Lorentz transformation. In order to obtain the solutions, invariant under such a transformation, of the Klein-Gordon equation

$$(\Box - m^2)f(x) = -\delta(x), \tag{1.1}$$

we follow the physicists by applying to (1.1) a slightly modified Fourier transformation F^*. This transformation applied to a function of the class $L(-\infty, +\infty)$ is defined by

$$F^*[f(x)] = \hat{f}(k) = \int_{-\infty}^{+\infty} e^{ik\cdot x} f(x)\, dx, \tag{2.2}$$

with $k\cdot x = k_0x_0 - k_1x_1 - k_2x_2 - k_3x_3$.

It is not difficult to show that for every Lorentz transformation Λ and for every $f(x) \in L(-\infty, +\infty)$ the following relation holds:

$$F^*[f(\Lambda x)] = \hat{f}(\Lambda k). \tag{2.3}$$

Using the transformation rules for distributions and (2.3) we obtain for every tempered distribution $f(x) \in S'$:

$$\begin{aligned}\langle F^*[f(\Lambda x)], \hat{\phi}(k)\rangle &= (2\pi)^4\langle f(\Lambda x), \phi(x)\rangle = (2\pi)^4\langle f(x), \phi(\Lambda^{-1}x)\rangle \\ &= \langle \hat{f}(k), \hat{\phi}(\Lambda^{-1}k)\rangle = \langle \hat{f}(\Lambda k), \hat{\phi}(k)\rangle.\end{aligned}$$

It follows immediately that (2.3) is also valid for distributions, and hence the Fourier transform F^* preserves Lorentz-invariance. The same is of course also true for the inverse transform F^{*-1}. The connection between the Fourier transform F^* and the usual one, viz.,

$$F[f(k)] = \int_{-\infty}^{+\infty} e^{i(x_0k_0+x_1k_1+x_2k_2+x_3k_3)} f(k)\, dk, \tag{2.4}$$

is given by the formula

$$FF^*[f(x_1, x_2, x_3, x_0)] = (2\pi)^4 f(x_1, x_2, x_3, -x_0), \tag{2.5}$$

which is equally valid for integrable functions as well distributions. We apply now F^* to (1.1) and we get

$$(m^2 - k^2)\hat{f}(k) = 1. \tag{2.6}$$

Since F^* and F^{*-1} preserve Lorentz-invariance, we have only to determine the Lorentz-invariant solutions of (2.6) and to transform the results again to configuration space.

The general Lorentz-invariant solution of (2.6) is readily obtained. A particular Lorentz-invariant solution of (2.6) is given by

$$\hat{f}_p(k) = \frac{1}{m^2 - k^2}, \tag{2.7}$$

where the distribution $(m^2 - k^2)^{-1}$ is defined as the Cauchy principal value,

$$\left\langle \frac{1}{m^2 - k^2}, \phi(k) \right\rangle = \lim_{\epsilon \to +0} \int_{|m^2-k^2|>\epsilon} \frac{\hat{\phi}(k)}{m^2 - k^2}\, dk. \tag{2.8}$$

The general solution $\hat{f}_H(k)$ of the homogeneous equation

$$(m^2 - k^2)\hat{f}_H(k) = 0 \tag{2.9}$$

is concentrated on the hyperboloid $(m^2 - k^2) = 0$.

In §3 we shall deal shortly with distributions concentrated on surfaces, and we shall show that, due to the postulated Lorentz-invariance,

$$\hat{f}_H(k) = c_+\delta_+(m^2 - k^2) + c_-\delta_-(m^2 - k^2), \tag{2.10}$$

where $\delta_\pm(m^2 - k^2)$ are distributions concentrated on, respectively, the upper and lower sheets of the hyperboloid $m^2 - k^2 = 0$, and where c_+ and c_- are arbitrary constants.

The distributions $\delta_\pm(m^2 - k^2)$ are defined as

$$\begin{aligned} &\langle \delta_\pm(m^2 - k^2), \hat{\phi}(k) \rangle \\ &= \frac{1}{2}\int_0^\infty \int_\Omega (\kappa^2 + m^2)^{-1/2} \kappa^2\, \hat{\phi}(\kappa\omega_1, \kappa\omega_2, \kappa\omega_3, \pm\sqrt{\kappa^2 + m^2})\, d\kappa\, d\Omega \\ &= \frac{1}{2}\int_0^\infty (\kappa^2 + m^2)^{-1/2} \kappa^2 \bar{\hat{\phi}}(\kappa, \pm\sqrt{\kappa^2 + m^2})\, d\kappa, \end{aligned} \tag{2.11}$$

where $\kappa^2 = k_1^2 + k_2^2 + k_3^2$, Ω denotes the unit sphere in (k_1, k_2, k_3)-space and $d\Omega$ its surface measure. $\bar{\hat{\phi}}(\kappa, k_0)$ is apart from a constant the mean value of $\hat{\phi}(k)$ on a sphere with radius κ in R_3 [8, Chap. I, §7]. Moreover, we have the relation

$$(2.12) \qquad \delta(m^2 - k^2) = \delta_+(m^2 - k^2) + \delta_-(m^2 - k^2).$$

Hence it follows from (2.7) and (2.10) that the general Lorentz-invariant solution of (2.6) is given by

$$(2.13) \quad \hat{f}(k) = (m^2 - k^2)^{-1} + c_+\delta_+(m^2 - k^2) + c_-\delta_-(m^2 - k^2).$$

All that remains is to determine the inverse transforms of $(m^2 - k^2)^{-1}$ and $\delta_\pm(m^2 - k^2)$. According to (2.5) we have to calculate the usual Fourier transforms of these distributions and to change consecutively x_0 to $-x_0$ and to divide by $(2\pi)^4$.

For the determination of the Fourier transform of (2.13) we use the following important formula:

$$(2.14) \qquad \begin{aligned} \frac{1}{m^2 - k^2 \pm i0} &\triangleq \lim_{\epsilon \to +0} \frac{1}{(m^2 - k^2) \pm i\epsilon(k_1^2 + k_2^2 + k_3^2 + k_0^2)} \\ &= \frac{1}{m^2 - k^2} \mp i\pi\delta(m^2 - k^2). \end{aligned}$$

(We use $\triangleq$ to denote "equals by definition".) This formula, being a generalization of the well-known one-dimensional equation

$$(2.15) \qquad \lim_{\epsilon \to +0} \frac{1}{x \pm i\epsilon} = \frac{1}{x} \mp i\pi\delta(x),$$

will be proved in §4.

Gelfand and Shilov have already given the Fourier transforms of $(m^2 - k^2 \pm i0)^{-1}$ (see [5, Chap. III, §2.8]). Using this result we obtain, with the aid of (2.14), $F[(m^2 - k^2)^{-1}]$ and $F[\delta(m^2 - k^2)]$ by addition and subtraction.

The only final problem is to determine $F[\delta_+(m^2 - k^2)]$ and $F[\delta_-(m^2 - k^2)]$. This will be done rigorously in §5, but we give here a sketch of the method which will be used. Let us suppose for the moment that

$$F[\delta(m^2 - k^2)] = X(x_1, x_2, x_3, x_0);$$

then it will be shown that it is possible to make the Hilbert-splitting

$$X(x_1, x_2, x_3, x_0) = X_1(x_1, x_2, x_3, x_0) + X_2(x_1, x_2, x_3, x_0),$$

where X_1 and X_2 can be continued analytically into the upper and lower halves, respectively, of the complex $(x_0 + iy_0)$-plane. According to a theorem

that the distributional limits $g(x_1, \cdots, x_n, x_0 \pm i0)$ of a function $g(x_1, \cdots, x_n, x_0 + iy_0)$ holomorphic in the upper or lower half-plane $y_0 > 0$, respectively, $y_0 < 0$, are Fourier transforms of distributions concentrated in the regions $k_0 \geqq 0$, respectively, $k_0 \leqq 0$, one finds after some reasoning that

$$F[\delta_+(m^2 - k^2)] = \lim_{y_0 \to +0} X_1(x_1, x_2, x_3, x_0 + iy_0)$$

and

$$F[\delta_-(m^2 - k^2)] = \lim_{y_0 \to -0} X_2(x_1, x_2, x_3, x_0 + iy_0).$$

Finally we present in §6 four important special Lorentz-invariant solutions of the differential equation (1.1), namely, the causal Green's functions $\Delta_C(x)$ and $\Delta_{AC}(x)$ and the retarded and advanced Green's functions $\Delta_R(x)$, respectively, $\Delta_A(x)$.

3. The general Lorentz-invariant solution of $(m^2 - k^2)\hat{f}(k) = 1$. The general Lorentz-invariant solution of the equation

$$(m^2 - k^2)\hat{f}(k) = 1 \tag{2.6}$$

can be obtained as the sum of a particular Lorentz-invariant solution $\hat{f}_p(k)$ and the general Lorentz-invariant solution $\hat{f}_H(k)$ of the homogeneous equation

$$(m^2 - k^2)\hat{f}_H(k) = 0. \tag{2.9}$$

The general solution of (2.9) is clearly concentrated on the hyperboloid $m^2 - k^2 = 0$, and hence we have to deal with distributions concentrated on surfaces. We introduce this distribution using an elegant method due to R. T. Seeley [14].

The distribution $\theta(P)$ is defined as

$$\langle \theta(P), \phi(x) \rangle = \int_{P \geqq 0} \phi(x)\, dx, \tag{3.1}$$

where $P(x_1, x_2, \cdots, x_n) = 0$ is some surface in R_n and P is a C^∞ function with $\nabla P = (\partial P/\partial x_1, \partial P/\partial x_2, \cdots, \partial P/\partial x_n)$ nowhere zero on $\{P = 0\}$. The distribution $\delta(P)$ is now introduced with the aid of the distribution $\theta(P)$, viz.,

$$\begin{aligned} \langle \delta(P), \phi(x) \rangle &= \lim_{c \to 0} \frac{1}{c} \langle \theta(P + c) - \theta(P), \phi(x) \rangle \\ &= \lim_{c \to 0} \frac{1}{c} \int_{-c \leqq P < 0} \phi(x)\, dx_1\, dx_2 \cdots dx_n. \end{aligned} \tag{3.2}$$

The existence of this limit presents no difficulty since we may write for the latter integral

$$(3.3)\quad \langle\delta(P), \phi(x)\rangle = \lim_{c\to 0} \frac{1}{c}\int_{P=0} \phi c \frac{d\sigma}{|\nabla P|} = \int_{P=0} \phi(x_1, \cdots, x_n) \frac{d\sigma}{|\nabla P|},$$

where $d\sigma$ is the surface measure on $\{P = 0\}$ and $|\nabla P| = \sqrt{(\nabla P, \nabla P)} \neq 0$; $dx_1\, dx_2 \cdots dx_n = c\,d\sigma/|\nabla P|$.

The derivatives $\delta^{(k)}(P)$ of the distribution $\delta(P)$ are defined by the rule

$$(3.4)\quad \delta^{(k+1)}(P) = \lim_{c\to 0} \frac{1}{c}[\delta^{(k)}(P + c) - \delta^{(k)}(P)], \qquad k = 0, 1, 2 \cdots.$$

It follows from (3.2) or (3.3) that $P\delta(P) = 0$; repeated differentiation of this equation with respect to P yields the useful formula

$$(3.5)\qquad P\delta^{(k)}(P) = -k\delta^{(k-1)}(P), \qquad k = 0, 1, 2, \cdots.$$

The derivative of $\delta^{(k)}(P)$ with respect to one of the independent variables x_i is given by the chain rule:

$$(3.6)\qquad \frac{\partial}{\partial x_i}\delta^{(k)}(P) = \delta^{(k+1)}(P)\frac{\partial P}{\partial x_i}.$$

As a useful example we give now the distribution $\delta(m^2 - k^2)$, which is concentrated on the hyperboloid $m^2 + k_1^2 + k_2^2 + k_3^2 - k_0^2 = 0$. Introducing spherical space coordinates

$$k_i = \kappa\omega_i, \quad i = 1, 2, 3, \quad \text{with} \quad \kappa = (k_1^2 + k_2^2 + k_3^2)^{1/2},$$

and taking instead of k_0 the new variable

$$P = m^2 + \kappa^2 - k_0^2,$$

we obtain

$$\begin{aligned}
\langle\delta(m^2 - k^2), \phi(k)\rangle &= \langle\delta(P), \hat\phi(k)\rangle \\
&= \lim_{c\to 0}\Bigg[\frac{1}{2c}\int_{-c\leq P<0} (m^2 + \kappa^2 - P)^{-1/2}\kappa^2\hat\phi(\kappa\omega_1, \kappa\omega_2, \kappa\omega_3, \\
&\qquad\qquad + \sqrt{m^2 + \kappa^2 - P})\, d\kappa\, d\Omega\, dP \\
&\quad + \frac{1}{2c}\int_{-c\leq P<0} (m^2 + \kappa^2 - P)^{-1/2}\kappa^2\hat\phi(\kappa\omega_1, \kappa\omega_2, \kappa\omega_3, \\
&\qquad\qquad - \sqrt{m^2 + \kappa^2 - P})\, d\kappa\, d\Omega\, dP\Bigg] \\
&= \frac{1}{2}\int_0^\infty\int_\Omega (m^2 + \kappa^2)^{-1/2}\kappa^2\hat\phi(\kappa\omega_1, \kappa\omega_2, \kappa\omega_3, + \sqrt{m^2 + \kappa^2})\, d\kappa\, d\Omega \\
&\quad + \frac{1}{2}\int_0^\infty\int_\Omega (m^2 + \kappa^2)^{-1/2}\kappa^2\hat\phi(\kappa\omega_1, \kappa\omega_2, \kappa\omega_3, - \sqrt{m^2 + \kappa^2})\, d\kappa\, d\Omega.
\end{aligned}$$

Performing the integration over the unit sphere Ω in (k_1, k_2, k_3)-space we get

$$(3.7)\quad \langle \delta(m^2 - k^2), \hat{\phi}(k) \rangle = \frac{1}{2} \int_0^\infty (m^2 + \kappa^2)^{-1/2} \kappa^2 \{ \bar{\hat{\phi}}(\kappa, + \sqrt{m^2 + \kappa^2}) + \bar{\hat{\phi}}(\kappa, - \sqrt{m^2 + \kappa^2}) \} \, d\kappa,$$

where $\bar{\hat{\phi}}(\kappa, k_0)$ is apart from a constant the mean value of $\hat{\phi}(k)$ over a sphere with radius κ in R_3.

The distribution $\delta(m^2 - k^2)$ is concentrated on the surface of the hyperboloid $m^2 - k^2 = 0$. The distributions $\delta_\pm(m^2 - k^2)$ are defined by the parts of $\delta(m^2 - k^2)$ which are concentrated on, respectively, the upper and lower sheets of the hyperboloid $m^2 - k^2 = 0$; hence we have the formulas

$$(3.8)\quad \langle \delta_+(m^2 - k^2), \hat{\phi}(k) \rangle = \frac{1}{2} \int_0^\infty (m^2 + \kappa^2)^{-1/2} \kappa^2 \bar{\hat{\phi}}(\kappa, + \sqrt{m^2 + \kappa^2}) \, d\kappa$$

and

$$(3.9)\quad \langle \delta_-(m^2 - k^2), \hat{\phi}(k) \rangle = \frac{1}{2} \int_0^\infty (m^2 + \kappa^2)^{-1/2} \kappa^2 \bar{\hat{\phi}}(\kappa, - \sqrt{m^2 + \kappa^2}) \, d\kappa.$$

It follows from (3.3) that $\delta_\pm(P)$ is even in P. Taking the limit for $m \to 0$ we obtain the distributions $\delta_\pm(k^2)$ concentrated on, respectively, the forward and backward lightcones; they are given by the expressions

$$(3.10)\quad \langle \delta_\pm(k^2), \hat{\phi}(k) \rangle = \frac{1}{2} \int_0^\infty \kappa \bar{\hat{\phi}}(\kappa, \pm \kappa) \, d\kappa.$$

We return now to the solution of the homogeneous equation

$$(2.9)\quad (m^2 - k^2) \hat{f}_H(k) = 0.$$

Because the only Lorentz-invariant subsets of the hyperboloid $m^2 - k^2 = 0$ are its upper and lower sheets, the general Lorentz-invariant solution of (2.9) consists of one invariant distribution concentrated on the upper sheet and one invariant distribution concentrated on the lower sheet.

In the same way as for distributions in one independent variable one can show that a distribution concentrated on a surface $P = 0$ is a linear combination of $\delta(P)$ and its derivatives $\delta^{(k)}(P)$. Because of the relations (2.9) and (3.5) the only distributions which are to be considered for a Lorentz-invariant solution of (2.9) are $\delta_+(m^2 - k^2)$ and $\delta_-(m^2 - k^2)$; hence we obtain

$$(2.10)\quad \hat{f}_H(k) = c_+ \delta_+(m^2 - k^2) + c_- \delta_-(m^2 - k^2).$$

Due to the required Lorentz-invariance, the coefficients c_+ and c_- must be constant.

Combining the results (2.7) and (2.10) we get for the general Lorentz-invariant solution of the equation

$$(m^2 - k^2)\hat{f}(k) = 1$$

the result

(2.13) $\hat{f}(k) = (m^2 - k^2)^{-1} + c_+\delta_+(m^2 - k^2) + c_-\delta_-(m^2 - k^2).$

4. The Fourier transforms of $(m^2 - k^2)^{-1}$ and $\delta(m^2 - k^2)$. In order to obtain the Fourier transforms of $(m^2 - k^2)^{-1}$ and $\delta(m^2 - k^2)$ we shall use the formula

(2.14)
$$\frac{1}{(m^2 - k^2) \pm i0} \triangleq \lim_{\epsilon \to +0} \frac{1}{(m^2 - k^2) \pm i\epsilon(k_1^2 + k_2^2 + k_3^2 + k_0^2)} = \frac{1}{m^2 - k^2} \mp i\pi\delta(m^2 - k^2).$$

This formula is proved as follows.

Proof. We introduce the differential operator

(4.1)
$$L_\epsilon = \frac{1}{1 \pm i\epsilon}\left(\frac{\partial^2}{\partial x_1^2} + \frac{\partial^2}{\partial x_2^2} + \frac{\partial^2}{\partial x_3^2}\right) - \frac{1}{1 \mp i\epsilon}\frac{\partial^2}{\partial x_0^2}$$

with

$$\lim_{\epsilon \to +0} L_\epsilon = \Box.$$

Putting

(4.2) $m^2 - x^2 \pm i\epsilon(x_1^2 + x_2^2 + x_3^2 + x_0^2) = m^2 - x^2 \pm i\epsilon(x, x) = \mathcal{P}_\epsilon,$

we obtain for all values of λ, ϵ and m not zero, the relation

$$L_\epsilon[\mathcal{P}_\epsilon^{\lambda+1}] = 4(\lambda + 1)(\lambda + 2)\mathcal{P}_\epsilon^{\lambda} - 4m^2\lambda(\lambda + 1)\mathcal{P}_\epsilon^{\lambda-1},$$

or, for $\lambda \neq -1$,

$$\mathcal{P}_\epsilon^{-\lambda} L_\epsilon\left[\frac{\mathcal{P}_\epsilon^{\lambda+1}}{\lambda + 1}\right] = \mathcal{P}_\epsilon^{-\lambda} L_\epsilon\left[\frac{\exp[(\lambda + 1)\log \mathcal{P}_\epsilon]}{\lambda + 1}\right] = 4(\lambda + 2) - 4m^2\lambda\mathcal{P}_\epsilon^{-1}.$$

After expanding the exponential function into a power series and taking the limit for $\lambda \to -1$ we get the identity

(4.3)
$$\mathcal{P}_\epsilon L_\epsilon[\log \mathcal{P}_\epsilon] = 4m^2\mathcal{P}_\epsilon^{-1} + 4,$$

and thus

(4.4)
$$\lim_{\epsilon \to +0} \mathcal{P}_\epsilon^{-1} = \lim_{\epsilon \to +0}\left\{\frac{\mathcal{P}_\epsilon}{4m^2} L_\epsilon[\log \mathcal{P}_\epsilon]\right\} - \frac{1}{m^2}.$$

The right-hand side of (4.4) may now be reduced as follows:

$$\lim_{\epsilon\to+0}\left\{\frac{\mathcal{P}_\epsilon}{4m^2}L_\epsilon[\log\mathcal{P}_\epsilon]\right\}-\frac{1}{m^2}=\frac{m^2-x^2}{4m^2}\,\Box\,[\log|m^2-x^2|]$$

$$+\frac{m^2-x^2}{4m^2}\,\Box\,[\pm i\pi\theta(x^2-m^2)]-\frac{1}{m^2}$$

$$=\frac{m^2-x^2}{4m^2}\,\Box\,[\log|m^2-x^2|]$$

$$-\frac{1}{m^2}\mp i\pi\delta(m^2-x^2);$$

and therefore we obtain

$$\text{(4.5)}\quad \lim_{\epsilon\to+0}\mathcal{P}_\epsilon^{-1}=\frac{m^2-x^2}{4m^2}\,\Box\,[\log|m^2-x^2|]-\frac{1}{m^2}\mp i\pi\delta(m^2-x^2).$$

The distribution $\langle[(m^2-x^2)/4m^2]\Box[\log|m^2-x^2|],\phi(x)\rangle$ may be written as:

$$\left\langle\frac{m^2-x^2}{4m^2}\,\Box\,[\log|m^2-x^2|],\phi(x)\right\rangle$$

$$=\frac{1}{4m^2}\langle\log|m^2-x^2|,\Box[(m^2-x^2)\phi(x)]\rangle$$

$$=\lim_{\delta\to0}\frac{1}{4m^2}\int_{|m^2-x^2|>\delta}\log|m^2-x^2|\cdot\Box[(m^2-x^2)\phi(x)]\,dx.$$

Applying Green's theorem to the right-hand side of this equation and observing that the integrals over the surfaces $m^2-x^2=\pm\delta$ cancel each other in the limit $\delta\to0$, we get

$$\left\langle\frac{m^2-x^2}{4m^2}\,\Box\,[\log|m^2-x^2|],\phi(x)\right\rangle$$

$$=\lim_{\delta\to0}\frac{1}{4m^2}\int_{|m^2-x^2|>\delta}(m^2-x^2)\,\Box\,[\log|m^2-x^2|]\cdot\phi(x)\,dx.$$

Calculating finally $(m^2-x^2)\Box[\log|m^2-x^2|]$ we obtain

$$\left\langle\frac{m^2-x^2}{4m^2}\,\Box\,[\log|m^2-x^2|],\phi(x)\right\rangle=\lim_{\delta\to0}\int_{|m^2-x^2|>\delta}\frac{\phi(x)\,dx}{m^2-x^2}+\int\frac{\phi(x)\,dx}{m^2}.$$

Substituting this result into the right-hand side of (4.5) we have

$$\text{(4.6)}\quad \lim_{\epsilon\to+0}\mathcal{P}_\epsilon^{-1}=(m^2-x^2\pm i0)^{-1}=(m^2-x^2)^{-1}\mp i\pi\delta(m^2-x^2),$$

where $(m^2-x^2)^{-1}$ should be taken in the sense of a Cauchy principal value.

The result (4.6) is also valid for $m = 0$; the proof is quite similar to the one given above. For $m \neq 0$, (4.6) may be generalized to an arbitrary number of variables and to arbitrary negative integer values of λ (see [15]). Following the method of Gelfand and Shilov [5, Chap. III, §2.8] we now determine the Fourier transforms of $\{(m^2 - k^2) \pm i0\}^{-1}$, from the result of which those of $(m^2 - k^2)^{-1}$ and $\delta(m^2 - k^2)$ immediately follow by aid of the formula (4.6) (or (2.14)).

If D is an arbitrary positive definite quadratic form in k, viz.,

$$D = (k, Gk) = \sum_{r,s=0}^{3} g_{rs} k_r k_s \,, \tag{4.7}$$

where G denotes the matrix of the coefficients g_{rs} and

$$(k, l) = k_1 l_1 + k_2 l_2 + k_3 l_3 + k_0 l_0 \,,$$

it is not difficult to calculate the Fourier transform of $(m^2 + D)^\lambda$ with $\operatorname{Re} \lambda < -2$. The result is:

$$F[(m^2 + D)^\lambda] = \frac{2\pi^2}{|G|^{1/2}} \left(\frac{2m}{E^{1/2}}\right)^{\lambda+2} \frac{K_{\lambda+2}(mE^{1/2})}{\Gamma(-\lambda)} \,; \tag{4.8}$$

$|G|$ is the determinant of the matrix G, and E is the positive definite quadratic form $\sum_{r,s=0}^{3} g^{rs} x_r x_s$, the coefficients of which form the inverse matrix of G. $K_{\lambda+2}$ is the modified Bessel function of order $\lambda + 2$ and the square roots are taken positive.

Performing analytical continuations with respect to λ and with respect to the coefficients g_{rs} into those ranges of complex values of g_{rs} where the so obtained new quadratic form $\mathfrak{D}$ has either a positive or negative definite imaginary part, one obtains for this complex quadratic form $\mathfrak{D}$ and for all values of λ the formula

$$F[(m^2 + \mathfrak{D})^\lambda] = \frac{2\pi^2}{|\mathcal{G}|^{1/2}} \left(\frac{2m}{\mathcal{E}^{1/2}}\right)^{\lambda+2} \frac{K_{\lambda+2}(m\mathcal{E}^{1/2})}{\Gamma(-\lambda)} \,; \tag{4.9}$$

$|\mathcal{G}|$ is again the determinant of the coefficients of $\mathfrak{D}$, and $\mathcal{E}$ is a quadratic form in x, the coefficients of which form a matrix being again the inverse of the matrix of the coefficients of $\mathfrak{D}$. The function $\mathcal{E}^{1/2}$, considered as a function defined in a complex $\mathcal{E}$-plane, has a cut along the negative real axis.

Taking for λ the value $\lambda = -1$, and for $\mathfrak{D}$ the quadratic form

$$\begin{aligned} \mathfrak{D} &= k_1^2 + k_2^2 + k_3^2 - k_0^2 \pm i\epsilon(k_1^2 + k_2^2 + k_3^2 + k_0^2) \\ &= -k^2 \pm i\epsilon(k, k), \end{aligned}$$

and considering finally the limit for $\epsilon \to +0$, one gets the result of Gelfand

and Shilov:

$$(4.10) \qquad F[m^2 - k^2 \pm i0)^{-1}] = \mp 4\pi^2 im \frac{K_1\{m(-x^2 \mp i0)^{1/2}\}}{(-x^2 \mp i0)^{1/2}},$$

with $-x^2 = x_1^2 + x_2^2 + x_3^2 - x_0^2$. For more detailed information on the results (4.8)–(4.10) the reader is referred to [5, Chap. III, §2.8] and [8, Chap. IV, §4]. It follows now immediately from the result (2.14) that

$$(4.11) \qquad \begin{aligned} &F\left[\frac{1}{m^2 - k^2}\right] \\ &\quad = 2\pi^2 im\left[\frac{K_1\{m(-x^2 + i0)^{1/2}\}}{(-x^2 + i0)^{1/2}} - \frac{K_1\{m(-x^2 - i0)^{1/2}\}}{(-x^2 - i0)^{1/2}}\right] \end{aligned}$$

and

$$(4.12) \qquad \begin{aligned} &F[\delta(m^2 - k^2)] \\ &\quad = 2\pi m\left[\frac{K_1\{m(-x^2 + i0)^{1/2}\}}{(-x^2 + i0)^{1/2}} + \frac{K_1\{m(-x^2 - i0)^{1/2}\}}{(-x^2 - i0)^{1/2}}\right]. \end{aligned}$$

Using the relations

$$K_1(ze^{i\pi/2}) = -\tfrac{1}{2}\pi H_1^{(2)}(z) = -\tfrac{1}{2}\pi J_1(z) + \tfrac{1}{2}\pi i Y_1(z),$$

$$K_1(ze^{-i\pi/2}) = -\tfrac{1}{2}\pi H_1^{(1)}(z) = -\tfrac{1}{2}\pi J_1(z) - \tfrac{1}{2}\pi i Y_1(z),$$

where $H_1^{(i)}$ is the Hankel and Y_1 the Neumann function, we get

$$F\left[\frac{1}{m^2 - k^2}\right] = 0 \quad \text{for} \quad x^2 < 0,$$

i.e., outside the lightcone, and

$$F\left[\frac{1}{m^2 - k^2}\right] = -2\pi^3 m \frac{J_1(m\sqrt{x^2})}{\sqrt{x^2}} \quad \text{for} \quad x^2 > 0,$$

i.e., inside the lightcone.

However, for x^2 in a neighborhood of $x^2 = 0$, from (4.11) we obtain

$$\begin{aligned} F\left[\frac{1}{m^2 - k^2}\right] &= 2\pi^2 i\left[\frac{1}{-x^2 + i0} + \frac{1}{x^2 + i0}\right] + O(\log|x^2|) \\ &= 4\pi^3\delta(x^2) + O(\log|x^2|). \end{aligned}$$

Summarizing these results we may write

$$(4.13) \qquad F\left[\frac{1}{m^2 - k^2}\right] = -2\pi^3 m\theta(x^2)\frac{J_1(m\sqrt{x^2})}{\sqrt{x^2}} + 4\pi^3\delta(x^2).$$

In the same way we can reduce (4.12):

$$F[\delta(m^2 - k^2)] = 4\pi m \frac{K_1(m\sqrt{-x^2})}{\sqrt{-x^2}} \quad \text{for} \quad x^2 < 0,$$

i.e., outside the lightcone, and

$$F[\delta(m^2 - k^2)] = i\pi^2 m \frac{H_1^{(2)}(m\sqrt{x^2}) - H_1^{(1)}(m\sqrt{x^2})}{\sqrt{x^2}} \quad \text{for} \quad x^2 > 0,$$

i.e., inside the lightcone. The square roots are to be taken positive. Further, it follows from (4.12) that, for small values of x^2,

$$(4.12) \qquad \begin{aligned} F[\delta(m^2 - k^2)] &= 2\pi\left[\frac{1}{-x^2 + i0} - \frac{1}{x^2 + i0}\right] + O(\log|x^2|) \\ &= -\frac{4\pi}{x^2} + O(\log|x^2|), \end{aligned}$$

where $1/x^2$ should be interpreted in the sense of Cauchy.

Finally, also these results may be summarized in one single formula, viz.,

$$(4.14) \qquad \begin{aligned} F[\delta(m^2 - k^2)] &= 4\pi m\theta(-x^2) \frac{K_1(m\sqrt{-x^2})}{\sqrt{-x^2}} \\ &\quad + i\pi^2 m\theta(x^2) \frac{H_1^{(2)}(m\sqrt{x^2}) - H_1^{(1)}(m\sqrt{x^2})}{\sqrt{x^2}} \end{aligned}$$

The results (4.13) and (4.14) yield the Fourier transforms of $(m^2 - k^2)^{-1}$ and $\delta(m^2 - k^2)$; there remains now to determine the Fourier transforms of the distributions $\delta_\pm(m^2 - k^2)$; this will be done in the next section.

5. The Fourier transform of $\delta_\pm(m^2 - k^2)$. In order to determine $F[\delta_\pm(m^2 - k^2)]$ we need first a lemma.

LEMMA. *Let*

$$g(u_1, u_2, \cdots, u_n, w_0) = g(u_1, u_2, \cdots, u_n, u_0 + iv_0)$$

be holomorphic in the upper half-plane $v_0 > 0$ for any set of real values of $(u_1, u_2, \cdots, u_n)$. The function $g(u_1, \cdots, u_n, w_0)$ can be majorized in any region $v_0 > \delta > 0$ as

$$|g(u_1, \cdots, u_n, u_0 + iv_0)| < C_\delta \prod_{i=1}^{n} (1 + u_i^2)^{p_i} |u_0 + iv_0|^{p_0},$$

where the p_i, $i = 0, \cdots, n$, are positive integers independent of δ. If $\lim_{v_0 \to +0} g(u_1, u_2, \cdots, u_n, u_0 + iv_0)$ exists in the distributional sense on *the space S of test functions, then*

$$\lim_{v_0 \to +0} g(u_1, \cdots, u_n, u_0 + iv_0)$$

is the $(n+1)$*-dimensional Fourier transform of a distribution* $f_+(x_1, x_2, \cdots, x_n, x_0)$ *belonging to* S' *and vanishing for* $x_0 < 0$.

An analogous result holds of course for functions $g(u_1, \cdots, u_n, w_0)$ holomorphic in the lower half-plane $v_0 < 0$. Mutatis mutandis one obtains that

$$\lim_{v_0 \to -0} g(u_1, \cdots, u_n, u_0 + iv_0)$$

is the $(n+1)$-dimensional Fourier transform of a distribution $f_-(x_1, \cdots, x_n, x_0)$ belonging to S' and vanishing for $x_0 > 0$.

Remark. H. A. Lauwerier [16] has shown that a function $g(u + iv)$, holomorphic in the upper half-plane $v > 0$ and bounded by a polynomial uniformly in every half-plane $v \geqq \delta > 0$, possesses for $v \to +0$ a distributional limit $g(u) \in Z'$ which is the Fourier transform of a distribution $f_+(x) \in D'$ vanishing for $x < 0$. (D' is the dual of the space of test functions belonging to C^∞ and with compact support.)

E. J. Beltrami and M. R. Wohlers [17] have proved the same theorem for distributions $f(x) \in S'$; however, in this case one needs an extra condition, namely, the existence of the distributional limit of $g(u + iv)$ in S', which is no longer a consequence of the data as stated in the theorem of Lauwerier. In our case we deal with distributions in more independent variables, and therefore we have to make a modification of the result of Beltrami and Wohlers.

Proof of the lemma. According to the assumptions there exist positive integers $p_1, p_2, \cdots, p_n, p_0$ such that

$$h(u_1, \cdots, u_n, u_0 + iv_0)$$
$$= \left\{ \prod_{i=1}^{n} (1 + u_i^2)^{-p_i - 1} \right\} (u_0 + iv_0)^{-p_0 - 2} g(u_1, \cdots, u_n, u_0 + iv_0)$$

is absolutely integrable over the whole space $R_{n+1}(u_1, \cdots, u_n, u_0)$ for any value of $v_0 > 0$. We consider now the integral

$$\frac{1}{(2\pi)^{n+1}} \int_{-\infty}^{+\infty} \cdots \int_{-\infty}^{+\infty} \exp\left[-i\{x_1u_1 + \cdots + x_nu_n + x_0(u_0 + iv_0)\}\right]$$
$$\cdot\, h(u_1, \cdots, u_n, u_0 + iv_0)\, du_1 \cdots du_n du_0$$

$$= \frac{1}{(2\pi)^{n+1}} \int_{-\infty}^{+\infty} \cdots \int_{-\infty}^{+\infty} \exp\left[-i(x_1u_1 + \cdots + x_nu_n)\right] du_1 \cdots du_n \tag{5.1}$$
$$\cdot \int_{-\infty}^{+\infty} \exp\left[-ix_0(u_0 + iv_0)\right] h(u_1, \cdots, u_n, u_0 + iv_0)\, du_0 .$$

Since h is holomorphic in the upper half-plane $v_0 > 0$, this integral is independent of v_0; hence the left-hand side of (5.1) is a function depending only on $x_1, \cdots, x_n, x_0$; taking $x_0 < 0$ and $v_0 \to +\infty$, it appears that this function, denoted by $f_+^*(x_1, \cdots, x_n, x_0)$, must vanish for $x_0 < 0$. It follows immediately from (5.1) that

$$F[e^{-v_0x_0}f_+^*(x_1, \cdots, x_n, x_0)] = h(u_1, \cdots, u_n, u_0 + iv_0).$$

Introducing the operator

$$D = \left\{\prod_{i=1}^{n}\left(1 - \frac{\partial^2}{\partial x_i^2}\right)^{p_i+1}\right\}\left(i\frac{\partial}{\partial x_0} + iv_0\right)^{p_0+2},$$

we obtain

$$F[De^{-v_0x_0}f_+^*(x_1, \cdots, x_n, x_0)] = g(u_1, \cdots, u_n, u_0 + iv_0).$$

Taking finally the limit for $v_0 \to +0$, which by supposition exists on S we obtain, due to the continuity of the Fourier transformation,

$$\lim_{v_0\to+0} g(u_1, u_2, \cdots, u_n, u_0 + iv_0) = F[\lim_{v_0\to+0} \{De^{-v_0x_0}f_+^*(x_1, \cdots, x_n, x_0)\}].$$

Finally, because f_+^* vanishes for $x_0 < 0$, we obtain the result that

$$\lim_{v_0\to+0} g(u_1, \cdots, u_n, u_0 + iv_0)$$

is the Fourier transform of a distribution vanishing for $x_0 < 0$. In the case that $g(u_1, \cdots, u_n, u_0 + iv_0)$ is holomorphic for $v_0 < 0$ one shows quite similarly that $g(u_1, \cdots, u_n, u_0 - i0)$ is the Fourier transform of a distribution vanishing for $x_0 > 0$.

We consider now the function

$$g(x_1, x_2, x_3, x_0) = 2\pi m \frac{K_1\{m\sqrt{-x^2}\}}{\sqrt{-x^2}} \tag{5.2}$$

with $x^2 < 0$ as a function of x_0, while x_1, x_2 and x_3 are supposed to have fixed values. The square root is taken positive. Putting

$$\sqrt{x_1^2 + x_2^2 + x_3^2} = r,$$

and introducing a complex $z_0 = x_0 + iy_0$ plane with cuts along the real axis, viz., $-\infty < x_0 < -r$ and $+r < x_0 < +\infty$, we can continue the function g analytically into the whole z_0-plane. Using the formulas

$$\begin{aligned} K_1(ze^{i\pi/2}) &= -\tfrac{1}{2}\pi H_1^{(2)}(z), \\ K_1(ze^{-i\pi/2}) &= -\tfrac{1}{2}\pi H_1^{(1)}(z), \end{aligned} \tag{5.3}$$

and the well-known expansions for K_1, $H_1^{(1)}$ and $H_2^{(2)}$ for small and large arguments [18], one finds after some simple considerations that the analytic continuation of $g(x_1, x_2, x_3, x_0)$ is uniformly bounded in any region y_0

$> \delta > 0$ and $y_0 < -\delta < 0$. Applying again (5.3) we obtain the following limits:

$$\lim_{y_0 \to +0} g(x_1, x_2, x_3, x_0 + iy_0) = \frac{-i\pi^2 m H_1^{(1)}(m\sqrt{x^2})}{\sqrt{x^2}} \quad \text{for} \quad x_0 > r,$$

$$\text{(5.4)} \quad \lim_{y_0 \to +0} g(x_1, x_2, x_3, x_0 + iy_0) = \frac{2\pi m K_1(m\sqrt{-x^2})}{\sqrt{-x^2}} \quad \text{for} \quad |x_0| < r,$$

$$\lim_{y_0 \to +0} g(x_1, x_2, x_3, x_0 + iy_0) = \frac{+i\pi^2 m H_1^{(2)}(m\sqrt{x^2})}{\sqrt{x^2}} \quad \text{for} \quad x_0 < -r,$$

and

$$\lim_{y_0 \to -0} g(x_1, x_2, x_3, x_0 + iy_0) = \frac{+i\pi^2 m H_1^{(2)}(m\sqrt{x^2})}{\sqrt{x^2}} \quad \text{for} \quad x_0 > r,$$

$$\text{(5.5)} \quad \lim_{y_0 \to -0} g(x_1, x_2, x_3, x_0 + iy_0) = \frac{2\pi m K_1(m\sqrt{-x^2})}{\sqrt{-x^2}} \quad \text{for} \quad |x_0| < r,$$

$$\lim_{y_0 \to -0} g(x_1, x_2, x_3, x_0 + iy_0) = \frac{-i\pi^2 m H_1^{(1)}(m\sqrt{x^2})}{\sqrt{x^2}} \quad \text{for} \quad x_0 < -r.$$

The square roots should of course again be taken as positive. We investigate now the limits of $g(x_1, x_2, x_3, x_0 + iy_0)$ for x_0 in the neighborhood of $x_0 = \pm r$.

For small values of $x_0 \mp r$ and y_0, the function $g(x_1, x_2, x_3, x_0 + iy_0)$ may be written as

$$g(x_1, x_2, x_3, x_0 + iy_0) = \frac{2\pi}{r^2 - (x_0 + iy_0)^2} + O\{\log | r^2 - (x_0 + iy_0)^2 |\},$$

and we have

$$\text{(5.6)} \quad \lim_{y_0 \to \pm 0} g(x_1, x_2, x_3, x_0 + iy_0) = \lim_{y_0 \to \pm 0} \frac{2\pi}{r^2 - (x_0 + iy_0)^2} + O\{\log |x^2|\}.$$

The limit of the first term is reduced as follows:

$$\text{(5.7)} \quad \begin{aligned} \lim_{y_0 \to \pm 0} \frac{2\pi}{r^2 - (x_0 + iy_0)^2} &= 2\pi \lim_{y_0 \to \pm 0} \frac{\partial}{\partial x_0} \left[\frac{1}{2r} \log \frac{r + (x_0 + iy_0)}{r - (x_0 + iy_0)} \right] \\ &= 2\pi \frac{\partial}{\partial x_0} \lim_{y_0 \to \pm 0} \left[\frac{1}{2r} \log \frac{r + (x_0 + iy_0)}{r - (x_0 + iy_0)} \right] \\ &= 2\pi \frac{\partial}{\partial x_0} \left[\frac{1}{2r} \log \left| \frac{r + x_0}{r - x_0} \right| \pm \frac{i\pi}{2r} \theta(x^2) \right] \\ &= \frac{-2\pi}{x^2} \pm 2i\pi^2 \frac{x_0}{r} \delta(x^2) \\ &= \frac{-2\pi}{x^2} \pm 2i\pi^2 \{\delta_+(x^2) - \delta_-(x^2)\}, \end{aligned}$$

where $1/x^2$ should be taken in the sense of Cauchy.

Inserting (5.7) into (5.6) we get for values of x_0 in the neighborhood of $x_0 = \pm r$,

$$\lim_{y_0 \to \pm 0} g(x_1, x_2, x_3, x_0 + iy_0)$$

$$(5.8) \qquad = \pm 2i\pi^2\{\delta_+(x^2) - \delta_-(x^2)\} - \frac{2\pi}{x^2} + O\{\log |x^2|\}.$$

The last two terms of this expression are already contained in (5.4) and (5.5); summarizing the results (5.4), (5.5) and (5.8) we obtain:

$$\lim_{y_0 \to +0} g(x_1, x_2, x_3, x_0 + iy_0)$$

$$(5.9) \qquad = \theta(x^2)\left[\theta(x_0)\frac{-i\pi^2 m H_1^{(1)}(m\sqrt{x^2})}{\sqrt{x^2}} + \theta(-x_0)\frac{i\pi^2 m H_1^{(2)}(m\sqrt{x^2})}{\sqrt{x^2}}\right]$$

$$+ \theta(-x^2)\frac{2\pi m K_1(m\sqrt{-x^2})}{\sqrt{-x^2}} + 2i\pi^2\{\delta_+(x^2) - \delta_-(x^2)\}$$

and

$$\lim_{y_0 \to -0} g(x_1, x_2, x_3, x_0 + iy_0)$$

$$(5.10) \qquad = \theta(x^2)\left[\theta(x_0)\frac{i\pi^2 m H_1^{(2)}(m\sqrt{x^2})}{\sqrt{x^2}} + \theta(-x_0)\frac{-i\pi^2 m H_1^{(1)}(m\sqrt{x^2})}{\sqrt{x^2}}\right]$$

$$+ \theta(-x^2)\frac{2\pi m K_1(m\sqrt{-x^2})}{\sqrt{-x^2}} - 2i\pi^2\{\delta_+(x^2) - \delta_-(x^2)\}.$$

The function $g(x_1, x_2, x_3, x_0 + iy_0)$ satisfies all conditions of the lemma, and hence it follows that

$$(5.11) \qquad g(x_1, x_2, x_3, x_0 \pm i0) = F[f_\pm(k_1, k_2, k_3, k_0)],$$

where f_+ and f_- are distributions vanishing for, respectively, $k_0 < 0$ and $k_0 > 0$.

Moreover, it follows from the result (4.14) that

$$g(x_1, x_2, x_3, x_0 + i0) + g(x_1, x_2, x_3, x_0 - i0) = F[\delta(m^2 - k^2)],$$

and hence

$$f_+(k) + f_-(k) = \delta(m^2 - k^2) = \delta_+(m^2 - k^2) + \delta_-(m^2 - k^2).$$

Because $\delta_+(m^2 - k^2)$ is concentrated on the upper sheet of the hyperboloid $m^2 - k^2 = 0$ and $\delta_-(m^2 - k^2)$ on its lower sheet, it follows further that

$$(5.12) \qquad \begin{aligned} f_+(k) &= \delta_+(m^2 - k^2) + h(k), \\ f_-(k) &= \delta_-(m^2 - k^2) - h(k), \end{aligned}$$

where $h(k)$ is a distribution concentrated on the coordinate plane $k_0 = 0$. According to (2.5) we have the relation

$$F^*[g(x_1, x_2, x_3, -x_0 + i0)] = (2\pi)^4 F^{-1}[g(x_1, x_2, x_3, x_0 + i0)] = (2\pi)^4 f_+(k_1, k_2, k_3, k_0).$$

The distribution $g(x_1, x_2, x_3, -x_0 + i0)$, and consequently also $f_+(k_1, k_2, k_3, k_0)$, are Lorentz-invariant; it follows from (5.12) that $h(k)$ is Lorentz-invariant. A distribution which is Lorentz-invariant and concentrated on the plane $k_0 = 0$ must be concentrated in the origin, and hence $h(k)$ is at most a Lorentz-invariant linear combination of $\delta(k_1, k_2, k_3, k_0)$ and its derivatives. Moreover, it follows from (5.9) and (5.10) that

$$g(x_1, x_2, x_3, -x_0 + i0) = g(x_1, x_2, x_3, +x_0 - i0),$$

and consequently,

$$f_+(k_1, k_2, k_3, -k_0) = f_-(k_1, k_2, k_3, +k_0).$$

Using formula (2.11) for $\delta_\pm(m^2 - k^2)$ it follows that $h(k)$ is odd in k_0.

Therefore, each term of $h(k)$ contains a derivative of $\delta(k)$ which is of odd order with respect to k_0; but such an expression is not properly Lorentz-invariant, and so it follows finally that $h(k) \equiv 0$ or

$$(5.13)\qquad \begin{aligned} f_+(k) &= \delta_+(m^2 - k^2), \\ f_-(k) &= \delta_-(m^2 - k^2). \end{aligned}$$

The combination of (5.9), (5.10), (5.11) and (5.13) gives the results

$$(5.14)\qquad \begin{aligned} F[\delta_+(m^2 - k^2)] = 2i\pi^2\{\delta_+(x^2) - \delta_-(x^2)\} &+ 2\pi m\theta(-x^2)\frac{K_1(m\sqrt{-x^2})}{\sqrt{-x^2}} \\ - i\pi^2 m\theta(x^2)\left[\theta(x_0)\frac{H_1^{(1)}(m\sqrt{x^2})}{\sqrt{x^2}} - \theta(-x_0)\frac{H_1^{(2)}(m\sqrt{x^2})}{\sqrt{x^2}}\right]&, \end{aligned}$$

$$(5.15)\qquad \begin{aligned} F[\delta_-(m^2 - k^2)] = -2i\pi^2\{\delta_+(x^2) - \delta_-(x^2)\} &+ 2\pi m\theta(-x^2)\frac{K_1(m\sqrt{-x^2})}{\sqrt{-x^2}} \\ + i\pi^2 m\theta(x^2)\left[\theta(x_0)\frac{H_1^{(2)}(m\sqrt{x^2})}{\sqrt{x^2}} - \theta(-x_0)\frac{H_1^{(1)}(m\sqrt{x^2})}{\sqrt{x^2}}\right]&. \end{aligned}$$

6. The Lorentz-invariant solutions of the Klein-Gordon equation. In §3 it has been shown that the Fourier transforms $F^*[f(x)]$ of the Lorentz-invariant solutions of the equation

$$(1.1) \qquad (\square - m^2)f(x) = -\delta(x)$$

are given by the general equation

$$(2.13) \qquad \hat{f}(k) = \frac{1}{m^2 - k^2} + c_+\delta_+(m^2 - k^2) + c_-\delta_-(m^2 - k^2).$$

In order to obtain $f(x)$ we have to apply to $\hat{f}(k)$ the inverse transformation F^{*-1}; according to (2.5) this can be done by applying first the Fourier-transform F and by changing consecutively x_0 into $-x_0$ and dividing by $(2\pi)^4$. The Fourier transforms F of the distributions

$$\frac{1}{m^2 - k^2} \quad \text{and} \quad \delta_\pm(m^2 - k^2)$$

have been determined in the last two sections, and so the problem to obtain the Lorentz-invariant solutions of the Klein-Gordon equation is solved. By specializing in $f(x)$ the values of the constants c_+ and c_-, one gets for $c_+ = c_- = \pi i$ the causal Green's function $\Delta_C(x)$, for $c_+ = c_- = -\pi i$ the anticausal Green's function $\Delta_{AC}(x)$, for $c_+ = -c_- = \pi i$ the retarded Green's function $\Delta_R(x)$, and finally for $c_+ = -c_- = -\pi i$ the advanced Green's function $\Delta_A(x)$.

For more information concerning these functions the reader is referred to [3], [8] and [13].

Elementary calculations give finally:

$$(6.1) \quad \Delta_C(x) = \frac{\delta(x^2)}{4\pi} - \frac{m}{8\pi}\theta(x^2)\frac{H_1^{(2)}(m\sqrt{x^2})}{\sqrt{x^2}} + \frac{im}{4\pi^2}\theta(-x^2)\frac{K_1(m\sqrt{-x^2})}{\sqrt{-x^2}},$$

$$(6.2) \quad \begin{aligned} \Delta_{AC}(x) = {} & \frac{\delta(x^2)}{4\pi} - \frac{m}{8\pi}\theta(x^2)\frac{H_1^{(1)}(m\sqrt{x^2})}{\sqrt{x^2}} \\ & - \frac{im}{4\pi^2}\theta(-x^2)\frac{K_1(m\sqrt{-x^2})}{\sqrt{-x^2}}, \end{aligned}$$

$$(6.3) \quad \begin{aligned} \Delta_R(x) &= \frac{1}{2\pi}\delta_+(x^2) - \frac{m}{4\pi}\theta(x^2)\frac{J_1(m\sqrt{x^2})}{\sqrt{x^2}} && \text{for } x_0 > 0, \\ \Delta_R(x) &= 0 && \text{for } x_0 < 0, \end{aligned}$$

and

$$(6.4) \quad \begin{aligned} \Delta_A(x) &= 0 && \text{for } x_0 > 0, \\ \Delta_A(x) &= \frac{1}{2\pi}\delta_-(x^2) - \frac{m}{4\pi}\theta(x^2)\frac{J_1(m\sqrt{x^2})}{\sqrt{x^2}} && \text{for } x_0 < 0. \end{aligned}$$

REFERENCES

[1] N. N. Bogoliubov and D. V. Shirkov, *Introduction to the Theory of Quantized Fields*, Interscience, New York, 1959.

[2] S. S. Schweber, H. A. Bethe and F. de Hoffman, *Mesons and Fields*, vol. I, Row, Peterson and Company, Evanston, Illinois, 1955.

[3] J. Hilgevoord, *Dispersion Relations and Causal Description*, North-Holland, Amsterdam, 1960.

[4] L. Schwartz, *Théorie des Distributions*, vols. I, II, Hermann, Paris, 1957, 1959.

[5] I. M. Gelfand and G. E. Shilov, *Verallgemeinerte Funktionen (Distributionen)*, vol. I, VEB. Deutscher Verlag der Wissenschaften, Berlin, 1960.

[6] H. Bremermann, *Distributions, Complex Variables, and Fourier Transforms*, Addison-Wesley, Reading, Massachusetts, 1965.

[7] A. H. Zemanian, *Distribution Theory and Transform Analysis*, McGraw-Hill, New York, 1965.

[8] E. M. de Jager, *Applications of distributions in mathematical physics*, Mathematical Center Tract 10, Amsterdam, 1964.

[9] P. D. Methée, *Sur les distributions invariantes dans le groupe des rotations de Lorentz*, Comment. Math. Helv., 28 (1954), pp. 225–269.

[10] ———, *L'équation des ondes avec second membre invariant*, Ibid., 32 (1957), pp. 153–164.

[11] L. Gårding and J. L. Lions, *Functional analysis*, Nuovo Cimento (10), 14 (1959), Supplement, §8, §9.

[12] J. Lavoine, *Solutions de l'équation de Klein-Gordon*, Bull. Sci. Math. (2), 85 (1961), pp. 57–72.

[13] E. M. de Jager, *The Lorentz-invariant solutions of the Klein-Gordon equation, I, II, III*, Nederl. Akad. Wetensch. Indag. Math., 25 (1963), pp. 515–531, 532–545, 546–558.

[14] R. T. Seeley, *Distributions on surfaces*, Report T.W. 78, Mathematical Center, Amsterdam, 1962.

[15] D. Bresters, *A note on distributions connected with quadratic forms*, Mathematical Communications of Twente Institute of Technology, (2), Enschede, The Netherlands, 1966.

[16] H. A. Lauwerier, *The Hilbert problem for generalized functions*, Arch. Rational Mech. Anal., 13 (1963), pp. 157–166.

[17] E. J. Beltrami and M. R. Wohlers, *Distributional boundary value theorems and Hilbert transforms*, Ibid., 18 (1965), pp. 304–309.

[18] H. Bateman, *Higher Transcendental Functions*, Bateman Manuscript Project, vol. II, McGraw-Hill, New York, 1965.

GENERALIZED FUNCTIONS IN ELEMENTARY PARTICLE PHYSICS AND PASSIVE SYSTEM THEORY: RECENT TRENDS AND PROBLEMS*

W. GÜTTINGER†

Abstract. The present state of our knowledge concerning the application of generalized functions to problems of passive system theory and elementary particle physics is reviewed. After a survey on the basic concepts of generalized functions some new developments are outlined: a compact integral representation for singular generalized functions is derived, leading to a generalization of the notion of pseudofunction and of Bochner's formula. An unsubtracted generalized Hilbert transform for increasing functions is discussed and confronted with subtracted dispersion relations obtained from boundary values of analytic functions. Product theories for generalized functions are developed and their connection with generalized Hilbert transforms is established. The local structure of infinite series of derivatives of delta functions is analyzed in terms of analytic functionals and a causal high frequency/energy bound in system theory is derived from this. Generalized functions with essential singularities arising from unstable systems are discussed. Linear and nonlinear passive system theories and elementary particle theories are outlined, and the new generalized function techniques are used to analyze various aspects of these theories. An outlook is given towards future developments.

Introduction. The mutual penetration of mathematical and physical thinking has been in the past of vital importance to the development of the exact sciences. It suffices to recall the case of quantum mechanics and the Hilbert space. Within the last decades, however, the gap between mathematics and physics has steadily widened and the impetus mathematics has gained from physics did not prevent it from becoming ever more abstract. This also happened to the theory of distributions and generalized functions which is now going the entangled paths of topology. However, the concept of generalized functions right now provides a rare opportunity to reunite physics and mathematics since it determines a new and more appropriate language in which to express the mathematical content of any particular physical theory.

Perfectly rigorous mathematical schemes do not always satisfy the physicist or the engineer. For, owing to the precise limitation between what is allowed and what is not, such schemes are often too narrow to give free

* Received by the editors November 3, 1966, and in revised form March 29, 1967. Presented by invitation at the Symposium on "The Applications of Generalized Functions" sponsored by the Air Force Office of Scientific Research at the 1966 Fall Meeting of Society for Industrial and Applied Mathematics held at the State University of New York at Stony Brook, September 12–14, 1966.

† Department of Physics, University of Munich, Munich, Germany, and Research Center and Department of Physics, New Mexico State University, Las Cruzes, New Mexico. Present address: University of Munich. This research was supported in part by the Deutsche Forschungsgemeinschaft.

play to formal entities and techniques, the success of which was uncontestable. It is therefore not surprising that new, strange but efficient objects and methods have been conceived by mathematically undisciplined spirits, mostly impatient theoretical physicists. Defended as a kind of shorthand or heuristic means for obtaining tentative solutions, such entities and techniques have become popular, e.g., Dirac's delta function, Hadamard's finite parts of divergent integrals and Heaviside's symbolic calculus. These conceptions are at the origin of what is now being called the theory of generalized functions.

The first one who realized again that physical necessities are always within the mathematical possibilities was L. Schwartz when he generalized classical analysis by his theory of distributions [1] to include singular functions and operations with them. Right now, however, distribution theory and its extensions [2], [3], [4], [5] have become too narrow to satisfy all the needs of the physicists. Thus the story starts anew: unpleasant subtractions in dispersion relations, essential singularities and unstable systems, broken symmetries and noncompact groups and all that are urging theoretical physicists to look for further generalizations of mathematical concepts.

The present paper is designed to review the present state of our knowledge concerning the application of generalized functions to physical systems and to sketch some new developments and techniques, part of them yet in virginal state, promising to become important tools for future research.

In §1 we review the basic concepts of generalized functions and operations on them. Section 2 is devoted to a survey on some recent developments. In §3 passive systems are discussed, while §4 is designed to analyze some concepts of quantum field theory. Finally, in §5, perspectives on future developments are indicated. For further details of part of the presented material we refer to [6], [7], [19].

1. Generalized functions: a review. When one asks for the properties of a physical system (a "black box"), one operates on it with a testing body and watches how the system responds to this stimulus. This response is a number $f\langle\varphi\rangle$ depending on both the object under investigation f and the testing body φ. By repeating the testing process with a set of different testing bodies, one obtains a corresponding set of numbers as responses, and both sets together characterize the object under investigation the better, the richer the class of testing bodies. For example, f may represent the density in space of a charge or probability distribution and $\varphi = \varphi(x)$ the density of a macroscopic testing body. Then, by the above testing process, the density of the object is represented by a functional f which assigns to every φ of a set ϕ a number $f\langle\varphi\rangle$. If, in particular, the density is a summable function $f(x)$ and the $\varphi(x)$ are summable too, the number representing the result of the testing of f by φ may be defined by the integral $f\langle\varphi\rangle = \int_{-\infty}^{\infty} f(x)\varphi(x)\,dx$. The point is that a knowledge of the class ϕ of testing functions $\varphi(x)$ and of the numbers $f\langle\varphi\rangle$ produced by them allows one to reconstruct the function $f(x)$ if the class of the φ is rich enough. Functionals of the above type are called generalized functions, but not all of them can be represented by simple integrals. The type of test functions to be used depends on the physical situation.

1.1. Test function spaces. The functions φ on which our functionals will act are called test functions, and the space ϕ formed by them is termed the test function space. By definition, ϕ is a linear topological space of functions $\varphi(x)$ defined on a set R of points x (e.g., $R = R^n$, C^n) and ϕ is a complete countably normed space. If the sequence $\varphi_n(x)$ converges within the topology of ϕ, then the sequence of numbers $\varphi_n(x_0)$ converges for any point $x_0 \in R$. The support of a continuous function $\varphi(x)$ is the closure of the set on which $\varphi(x) \neq 0$.

The most important test function spaces are the following ones:

(a) The space D of all infinitely differentiable complex-valued functions $\varphi(x)(x \in R^n)$ which vanish outside some finite and closed region, so that $(\varphi^{(n)}(x) = d^n\varphi/dx^n$, $x \in R^1)$

$$\varphi^{(n)}(\pm\infty) = 0. \tag{1.1}$$

A sequence $\varphi_\nu(x)$ is said to converge to zero in D if all these functions vanish outside one and the same bounded region and if they converge uniformly to zero together with their derivatives of any order.

(b) The space S of rapidly decreasing test functions consists of all infinitely differentiable functions $\varphi(x)$ which, together with all derivatives, approach zero more rapidly than any power of $1/|x|$ as $|x| \to \infty$. Thus, $\varphi(x) \in S$ satisfies an inequality $|x^n\varphi^{(m)}(x)| \leqq c_{nm}$, i.e., $\lim_{|x|\to\infty} |x^n\varphi^{(m)}(x)| = 0$ for any $n, m = 0, 1, 2, \cdots$. For the topology, cf. [1]–[6].

(c) Spaces of "type S" are intermediate between D and S and consist of functions φ satisfying inequalities of the type $\sup_x |x^n\varphi^{(m)}(x)| \leqq c_{nm}$, where (contrary to S) the c_{nm} are assumed to depend in a certain specific way on n and m. In particular, the space S_α^β is formed by all C^∞ functions φ satisfying $|x^n\varphi^{(m)}(x)| \leqq CA^nB^mn^{n\alpha}m^{m\beta}$, $\alpha \geqq 0$, $\beta \geqq 0$. If $\beta \leqq 1$, these $\varphi(x)$ can be analytically continued into the complex z-plane, $z = x + iy$, to yield analytic functions $\varphi(z)$ satisfying the inequalities $|\varphi(z)| \leqq C \exp[-a|x|^{1/\alpha} + b|y|^{1/(1-\beta)}]$.

(d) The space Z of slowly increasing entire analytic functions $\psi(z)$, $z = x + iy$, is made up of all those entire analytic functions ψ satisfying the inequalities $|z^n\psi(z)| \leqq c_n \exp(a|y|)$, where the constants c_n and a may depend on ψ. A sequence $\psi_\nu \in Z$ is said to converge to zero in Z if for each ψ_ν we have $|z^n\psi_\nu(z)| \leqq c_n \exp(a|y|)$ with c_n and a independent of ν and if these functions converge to zero uniformly on every interval of the real x-axis. In particular, the Taylor expansion $\psi(z) = \sum_0^\infty z^n\psi^{(n)}(0)/n!$ converges in Z and $\psi(z)|_{y=0} = \psi(x)$ is an element of S.

(e) The spaces D, S, S_α^β belong to a class of spaces $K\{M_p\}$, with φ satisfying $\|\varphi\|_p = \sup_{|q| \leqq |p|} M_p(x)|\varphi^{(q)}(x)|$, $p = 0, 1, 2, \cdots$. Here, the M_p, $1 \leqq M_1 \leqq M_2 \leqq \cdots$, are functions defined on R^n, and their values may be finite or infinite. The $M_p(x)$ are called the weights of the space. $M_p(x) = \sup_{|k| \leqq |p|} |x^k|$ yields $K\{M_p\} = S$. Similarly one defines the spaces $Z\{M_p\}$ with continuous weight functions $M_p(z)$, $0 < C(y) \leqq M_1 \leqq M_2 \leqq \cdots$, and $\psi \in Z\{M_p\}$ means that $\|\psi\|_p = \sup_z M_p(z)|\psi(z)|$, $1 \leqq p < \infty$. $M_p = \exp(-a|y|)\cdot\max_{|k| \leqq p} |z|^k$, $a = 1, 2, \cdots$, yields the space Z.

If ϕ is any test function space, then a function $a(x)$ which has the property that $\varphi \in \phi$ implies $a(x)\varphi(x) \in \phi$ and further that $\varphi_n \to 0$ implies $a\varphi_n \to 0$ is called a multiplier in the space ϕ. A multiplier in D is given by any infinitely differentiable function $a(x)$. For the multipliers in the above spaces, cf. [6].

The essential property of all ϕ considered is that their elements φ either are infinitely differentiable or analytic and that (1.1) is satisfied on the real axis.

1.2. Generalized functions. Suppose that by some rule we can associate with every $\varphi(x) \in \phi$ a complex number $f\langle\varphi\rangle$ (or (f, φ)), i.e., a functional, in such a way that the following conditions are satisfied: (i) the functional is linear, i.e., $f\langle\alpha_1\varphi_1 + \alpha_2\varphi_2\rangle = \alpha_1 f\langle\varphi_1\rangle + \alpha_2 f\langle\varphi_2\rangle$ for any two $\varphi_1, \varphi_2 \in \phi$ and for any two complex numbers α_1, α_2; (ii) the functional is continuous, i.e., if the sequence φ_n, $n = 1, 2, \cdots$, converges to zero in ϕ, $\varphi_n \to 0$, then the sequence of numbers $f\langle\varphi_n\rangle$ converges to zero, $f\langle\varphi_n\rangle \to 0$.

DEFINITION. A *generalized function* f is a linear and continuous functional on the space ϕ of test functions φ. The set of all generalized functions over a space ϕ will be denoted by ϕ'.

Generalized functions in D' are termed distributions [1], those in S' are called tempered distributions, while functionals in Z' or $Z'\{M_p\}$, i.e., analytic functionals, sometimes are called ultradistributions or hyperfunctions.

Let us consider some examples of the more common generalized functions:

(a) Let $f(x)$ be a locally summable function and let $\varphi \in D$. By means of $f(x)$ we may assign to every $\varphi \in D$ a number

$$f\langle\varphi\rangle = \int_{-\infty}^{\infty} f(x)\varphi(x)\,dx, \tag{1.2}$$

$f\langle\varphi\rangle$ being the value of the functional f at φ, and the integral in (1.2) going over the bounded region in R^n in which $\varphi(x)$ differs from zero. (dx or $d^n x$ denotes the volume element of R^n.)

(b) To form generalized functions of similar type on other spaces ϕ it is necessary to impose further restrictions on the growth of $f(x)$ as $|x| \to \infty$, ensuring the convergence of the integral (1.2) for all $\varphi \in \phi$. For example, for any $\varphi \in S$ it is sufficient to require that $f(x)$ should increase at infinity no more rapidly than some power of $|x|$.

Those generalized functions which can be represented in terms of a formula such as (1.2) are called regular; all others will be termed singular. The simplest regular generalized function is "the constant C", defined by $C\langle\varphi\rangle = \int_{-\infty}^{\infty} C\varphi(x)\,dx$. Another example is the step function,

$$\theta(x) = \begin{cases} 1 & \text{for } x > 0, \\ 0 & \text{for } x < 0, \end{cases}$$

with the associated generalized function being given by

$$\theta\langle\varphi\rangle = \int_0^{\infty} \varphi(x)\,dx. \tag{1.3}$$

Dirac's "delta function" is the best known singular generalized function, and is defined by

$$\delta\langle\varphi\rangle = \varphi(0). \tag{1.4}$$

This definition originates from the "classical", but meaningless, relation $\int_{-\infty}^{\infty} \delta(x)\varphi(x)\,dx = \varphi(0)$ (say, $\delta(x) = 0$ for $x \neq 0$, $\int_{-\infty}^{\infty} \delta(x)\,dx = 1$.)

Since a regular generalized function, defined by (1.2), allows one to reconstruct the generating locally summable function $f(x)$ almost everywhere, the regular generalized function f, (1.2), can be identified with the associated locally summable function $f(x)$. For this reason, and for calculation purposes, we shall use in what follows the notation $f(x)$ for both regular and singular generalized functions. When we write $f(x)$ for a generalized function f, we merely wish to indicate that f operates on test functions which depend on x. Thus we shall often denote $f\langle\varphi\rangle$ by $\int_{-\infty}^{\infty} f(x)\varphi(x)\,dx$, and write, e.g., $\delta\langle\varphi\rangle = \int_{-\infty}^{\infty} \delta(x)\varphi(x)\,dx = \varphi(0)$, although such symbolic notation is meaningless according to classical analysis. For example, we shall not cease to write symbolically the defining relation

$$\delta(x - x_0)\langle\varphi(x)\rangle = \varphi(x_0), \tag{1.5}$$

of which (1.4) is a particular case ($x_0 = 0$).

(c) Analytic functionals are constructed on the spaces Z, $Z\{M_p\}$. Confining ourselves to the space Z, we call a regular functional g any functional given by an expression of the form

$$g\langle\psi\rangle = \int_{-\infty}^{\infty} g(x)\psi(x)\,dx, \tag{1.6}$$

where $\psi(z) \in Z$, $z = x + iy$. The term analytic functional is used for functionals of the form

$$g\langle\psi\rangle = \int_{\Gamma} g(z)\psi(z)\,dz, \tag{1.7}$$

where $g(z)$ is a function and Γ some contour in the complex z-plane. For example, Dirac's delta function can be defined for complex arguments by $\delta\langle\psi\rangle = \psi(0)$ and by $\delta(z - z_0)\langle\psi(z)\rangle = \psi(z_0)$ for any complex z. Then, by Cauchy's formula,

$$\delta(z - z_0)\langle\psi(z)\rangle = \frac{1}{2\pi i}\int_{\Gamma_0} (z - z_0)^{-1}\psi(z)\,dz, \tag{1.8}$$

where Γ_0 is any closed contour encircling the point $z = z_0$ of the complex z-plane. Thus, $\delta(z)$ may be viewed as an analytic functional with the generating function $g = 1/2\pi iz$. Integration contours Γ_0 such that $\int_{\Gamma_0} \psi(z)\,dz = 0$ for any $\psi \in Z$ are called null contours. Two paths Γ_1 and Γ_2 are said to be equivalent if for any $\psi \in Z$ we have $\int_{\Gamma_1} \psi\,dz = \int_{\Gamma_2} \psi\,dz$.

(d) Representation theorems for generalized functions may be found in [1]–[6]. They show, loosely speaking, that a generalized function f admits a representation of the type $f = \sum_0^\infty \int_{-\infty}^{\infty} f_n(x)\varphi^{(n)}(x)\,dx$.

The support of a generalized function f is the smallest closed set of points outside of which f vanishes. f is said to vanish in the open neighborhood $U(x_0)$ of a point $x_0 \in R^n$ if $f\langle\varphi\rangle = 0$ for every φ which has its support in U. This definition applies to functionals defined on test functions over R^n. The support of an analytic functional will be discussed in §2.4. For example, $\delta(x - x_0)$ defined by (1.5) has the point $x = x_0$ as support. Comparison with (1.8) indicates that the support of analytic functionals related to null contours may be defined by the smallest closed set to which such a contour can be shrunk (to the point $z = z_0$ in (1.8)).

1.3. Operations with generalized functions. To find operations for generalized functions, we shall start from the rules for ordinary functions, translate them into functional language and then define the resulting formula to hold for any generalized function.

(i) Addition of two generalized functions f, g and multiplication by a complex number α are defined, respectively, by

$$(1.9)\qquad (f+g)\langle\varphi\rangle = f\langle\varphi\rangle + g\langle\varphi\rangle, \qquad (\alpha f)\langle\varphi\rangle = \alpha\cdot f\langle\varphi\rangle.$$

(ii) Multiplication of a generalized function f by a multiplier a of the space ϕ is defined by

$$(1.10)\qquad (a\cdot f)\langle\varphi\rangle = f\langle a\varphi\rangle.$$

For the notion of multiplier, see §1.1. The rule (1.10) originates from the trivial relation

$$(af)\langle\varphi\rangle = \int af\varphi\, dx = f\langle a\varphi\rangle,$$

valid for regular generalized functions in ϕ'. For example, let $f = \delta \in D'$ and let a be a multiplier in D, i.e., $a \in C^\infty$. Then the product $a\delta$, defined by (1.10), is given by

$$a(x)\cdot\delta(x) = a(0)\delta(x);$$

in particular, $x\cdot\delta(x) = 0$. For,

$$(a\delta)\langle\varphi\rangle = \delta\langle a\varphi\rangle = a(0)\varphi(0) = a(0)\delta\langle\varphi\rangle.$$

Products $f\cdot g$ of arbitrary generalized functions are discussed in §2.2 as well as in the subsequent sections.

(iii) The derivative f' of a generalized function f is defined by

$$(1.11)\qquad f'\langle\varphi\rangle = -f\langle\varphi'\rangle, \qquad x \in R^1, \quad \varphi' = d\varphi/dx.$$

This definition originates from the fact that if $f(x)$ has a continuous derivative $f'(x)$ we can associate with the latter the regular functional $f'\langle\varphi\rangle = \int_{-\infty}^{\infty} f'(x)\varphi(x)\, dx$ according to (1.2); and by integration by parts and using (1.1) we obtain in R^1,

$$f'(x)\langle\varphi\rangle = -\int_{-\infty}^{\infty} f(x)\varphi'(x)\, dx = -f(x)\langle\varphi'\rangle.$$

The derivative of $g \in Z'$ is defined by $g'\langle\psi\rangle = -g\langle\psi'\rangle$; partial derivatives are defined similarly, and repeated application of (1.11) yields $f^{(n)}\langle\varphi\rangle = (-1)^n f\langle\varphi^{(n)}\rangle$, i.e., every generalized function has derivatives of all orders. If a is a multiplier, then $(af)' = a'f + af'$.

Examples. $\theta'(x) = \delta(x)$, for, using (1.3), we have

$$\theta'\langle\varphi\rangle = -\theta\langle\varphi'\rangle = -\int_0^\infty \varphi'(x)\, dx = \varphi(0) = \delta\langle\varphi\rangle.$$

The nth derivative of $\delta(z - z_0)$ is given by

$$\delta^{(n)}(z - z_0)\langle\psi\rangle = (-1)^n\psi^{(n)}(z_0),$$

and one verifies $a(x)\delta'(x) = a(0)\delta'(x) - a'(0)\delta(x)$ by operating with both sides of this equation on $\varphi(x)$ using (1.10), (1.11).

(iv) Singular generalized functions are generated—and therefore defined—by differentiating regular generalized functions. Consider, e.g., the locally summable function

$$(1.12) \qquad x_+^\lambda = \begin{cases} x^\lambda & \text{for } x > 0, \qquad -1 < \operatorname{Re}\lambda < 0, \\ 0 & \text{for } x < 0 \end{cases}$$

(sometimes symbolically denoted by $x_+^\lambda = \theta(x)x^\lambda$ also if $\operatorname{Re}\lambda < -1$). The associated generalized function is given by $x_+^\lambda\langle\varphi\rangle = \int_0^\infty x^\lambda\varphi(x)\,dx$ in virtue of (1.2). Its derivative exists according to (1.11) and is given by

$$(x_+^\lambda)'\langle\varphi\rangle = -\int_0^\infty x^\lambda\varphi'\,dx = -\lim_{\epsilon\to0}\int_\epsilon^\infty x^\lambda\varphi'\,dx.$$

Integrating by parts and taking into account that $\varphi(\infty) = 0$ according to (1.1) we obtain (in virtue of $\epsilon^\lambda\varphi(\epsilon) = \epsilon^\lambda\varphi(0) + o(\epsilon)$, $-1 < \operatorname{Re}\lambda < 0$) the result ($-1 < \operatorname{Re}\lambda < 0$)

$$(1.13) \qquad \begin{aligned} (x_+^\lambda)'\langle\varphi\rangle &= \lim_{\epsilon\to0}\left[\lambda\int_\epsilon^\infty x^{\lambda-1}\varphi(x)\,dx + \epsilon^\lambda\varphi(0)\right] \\ &= \lambda\int_0^\infty x^{\lambda-1}[\varphi(x) - \varphi(0)]\,dx. \end{aligned}$$

Obviously, $(x_+^\lambda)'$ is no longer a regular functional, and the generalized function derivative of x_+^λ certainly is not equal to the "functional" $I = \int_0^\infty \lambda x^{\lambda-1}\varphi\,dx$ associated with the ordinary function derivative $\lambda x_+^{\lambda-1}$, for this derivative is not summable so that (1.2) cannot be applied (yielding $I = \infty$). The right-hand side of (1.13) is called "pseudofunction (Pf)" $Pf\,x_+^\lambda\langle\varphi\rangle$, and (1.1) reads

$$(1.14) \qquad (Pf\,(x_+^\lambda))'\langle\varphi\rangle = Pf\,(\lambda x_+^{\lambda-1})\langle\varphi\rangle, \qquad -1 < \operatorname{Re}\lambda < 0.$$

$Pf\,(\lambda x_+^{\lambda-1})$ is identical with the "finite part" [8] of the divergent integral $I_\epsilon = \int_\epsilon^\infty \lambda x^{\lambda-1}\varphi\,dx$ obtained by subtracting from I_ϵ the part $-\epsilon^\lambda\varphi(0)$ which diverges as $\epsilon \to 0$ ($-1 < \operatorname{Re}\lambda < 0$). The generalized function $Pf\,x_+^{\lambda-1}$ coincides with $x_+^{\lambda-1}$ for all $\varphi(x)$ vanishing at $x = 0$, i.e., symbolically,

$Pf\, x_+^{\lambda-1} = x_+^{\lambda-1}$ for $x \neq 0$. Continuing the differentiation one finds that

$$(1.15) \qquad (Pf\, x_+^{\lambda})' = Pf\,(\lambda x_+^{\lambda-1}), \qquad \lambda \neq -1, -2, \cdots,$$

where

$$(1.16) \qquad Pf\, x_+^{\lambda}\langle\varphi\rangle = \int_0^\infty x^\lambda \left[\varphi(x) - \sum_{\mu=0}^{N} x^\mu \varphi^{(\mu)}(0)/\mu!\right] dx, \qquad \lambda \neq -1, -2, \cdots,$$

$N = [-(\operatorname{Re}\lambda + 1)]$ being the largest nonnegative integer less than or equal to $-(\operatorname{Re}\lambda + 1)$. (If $\operatorname{Re}\lambda > -1$, the sum in (1.16) has to be dropped.) By means of the step function $\theta(x)$, one may write, e.g.,

$$Pf\, x_+^{-3/2} = \lim_{\epsilon \to 0}\, [x^{-3/2}\theta(x - \epsilon) - 2\epsilon^{-1/2}\delta(x)],$$

and one sees that $Pf\, x_+^{-3/2}$ becomes well defined because the subtracted "repulsive core" $2\delta(x)/\sqrt{\epsilon}$ just compensates the singularity the classical function $x_+^{-3/2}$ has at $x = 0$. Pseudofunctions are sometimes also called "regularizations" of classically divergent quantities. This is rather misleading since there is nothing to be regularized in defining singular generalized functions by differentiating regular ones. In what follows we shall drop the symbol Pf in $Pf\, x_+^{\lambda}$ unless otherwise stated.

Differentiating $(\log x)_+\langle\varphi\rangle$ according to (1.11) one finds $(\log x)_+' = Pf\, x_+^{-1}$, where $Pf\, x_+^{-1}$ is defined by

$$Pf\, x_+^{-1}\langle\varphi\rangle = \lim_{\epsilon \to 0} \left[\int_\epsilon^\infty x^{-1}\varphi(x)\, dx + \varphi(0) \log \epsilon\right].$$

Continuing differentiation one obtains $(\psi(n) = \Gamma'(n)/\Gamma(n))$

$$(1.17) \qquad \frac{(-1)^{n-1}}{(n-1)!}(\log x)_+^{(n)} = Pf\, x_+^{-n} + \frac{(-1)^n}{(n-1)!}\,\delta^{(n-1)}(x)\cdot[\psi(n) - \psi(1)],$$

where $Pf\, x_+^{-n}$ is given by $(n = 1, 2, \cdots)$

$$(1.18) \qquad Pf\, x_+^{-n}\langle\varphi\rangle = \int_0^\infty x^{-n}\left[\varphi(x) - \sum_{\mu=0}^{n-2} \frac{x^\mu \varphi^{(\mu)}(0)}{\mu!} - \theta(1 - x)\,\frac{x^{n-1}\varphi^{(n-1)}(0)}{(n-1)!}\right] dx.$$

This is usually done in the literature [1]–[5] but, as we shall show in §2.1, (1.18) is a bad and even physically unacceptable definition which requires a strong modification.

(v) A generalized function f_λ depending on a complex parameter λ varying in a domain Λ of the complex λ-plane is said to be an analytic generalized function with respect to λ if the derivative of $f_\lambda\langle\varphi\rangle$ with respect to λ exists for all φ. If f_λ is analytic in λ in some domain, analytic continuation out of that domain follows the usual rules. For example, the generalized function $x_+^\lambda\langle\varphi\rangle$ is analytic for $\operatorname{Re}\lambda > -1$. Consider the identity

$$\int_0^\infty x^\lambda\varphi(x)\,dx = \int_0^a x^\lambda\left[\varphi(x) - \sum_0^N \frac{x^\mu\varphi^{(\mu)}(0)}{\mu!}\right]dx \qquad (1.19)$$

$$+ \int_a^\infty x^\lambda\varphi(x)\,dx + \sum_0^N \frac{a^{\mu+1+\lambda}\varphi^{(\mu)}(0)}{\mu!(\mu - 1 + \lambda)},$$

valid for $\operatorname{Re}\lambda > -1$ ($a > 0$ being some positive constant) with $N = [-(\operatorname{Re}\lambda + 1)]$. The left-hand side is analytic for $\operatorname{Re}\lambda > -1$, the right-hand side exists for all $\lambda \neq -1, -2, \cdots$ and thus defines the analytic continuation of x_+^λ from $\operatorname{Re}\lambda > -1$ to the complex λ-plane except for the points $\lambda = -1, -2, \cdots$ where the right-hand side has simple poles. It is easily seen [6] that the result of this continuation coincides with what we obtained in (1.16) by differentiating x_+^λ, $\operatorname{Re}\lambda > -1$, i.e., $Pf\, x_+^\lambda = \text{A.C. } x_+^\lambda$ for $\lambda \neq -1, -2, \cdots$, A.C. denoting analytic continuation. Extensions and generalizations of these results will be discussed in §2.1.

The above arguments in (iv) and (v) also apply to $x_-^\lambda = |x|^\lambda$ for $x < 0$ and $x_-^\lambda = 0$ for $x > 0$, i.e., to $x_-^\lambda\langle\varphi\rangle$ defined by $x_-^\lambda\langle\varphi\rangle = x_+^\lambda\langle\varphi(-x)\rangle$. Defining $|x|^\lambda = x_+^\lambda + x_-^\lambda$, $\epsilon(x)|x|^\lambda = x_+^\lambda - x_-^\lambda$ and x^{-n} by $\{|x|^{-n}\ (n \text{ even}), \epsilon(x)|x|^{-n}\ (n \text{ odd})\}$, where $\epsilon(x) = \operatorname{sgn} x$, one obtains representations for $|x|^{-n}$, $x^{-n} = \text{p.v. } x^{-n}$, etc. The $x_\pm^\lambda$ have poles of first order $\lambda = -1, -2, \cdots$ with

$$\operatorname*{Res}_{\lambda=-n} x_\pm^\lambda = (-1)^{n-1}\delta^{(n-1)}(\pm x)/(n-1)!, \quad n = 1, 2, \cdots. \qquad (1.20)$$

Hence, for the so-called Riesz-distributions

$$R_\pm^\lambda = x_\pm^\lambda/\Gamma(\lambda + 1), \qquad (1.21)$$

we have

$$R_\pm^{-n} = \delta^{(n-1)}(\pm x). \qquad (1.22)$$

Equation (1.21) serves to define nonintegral derivatives of delta functions.

(vi) A series $\sum_0^\infty f_n$ of generalized functions f_n is said to converge in ϕ' to f, $f = \sum_0^\infty f_n$, if for every $\varphi \in \phi$, the numerical series $\sum_0^\infty f_n\langle\varphi\rangle$ converges to the number $f\langle\varphi\rangle$. For example, for any $g \in Z'$ we have $\sum_0^\infty z_0^n g^{(n)}(z)/n! = g(z + z_0)$.

(vii) The convolution $f * g$ of f and g is defined by

$$(f * g)\langle\varphi(x)\rangle = f(x)g(y)\langle\varphi(x + y)\rangle$$

whenever the right-hand side exists (e.g., if f has compact support).

(viii) The Fourier transform of a generalized function f is defined as follows: Let $\varphi(x) \in \phi$ have a Fourier transform $\psi(k) = F\varphi(x)$ in the ordinary sense (where $F\varphi = \int_{-\infty}^{\infty} \exp(ikx)\varphi\, dx$ denotes the Fourier operator, $x \in R^n$ and $k \in C^n = R_p{}^n \times R_q{}^n$, $k = p + iq$), and let f be a generalized function $f \in \phi'$ in x. Then the Fourier transform $\tilde{f}$ of f, $\tilde{f} = Ff$, is defined by the relation (in R^n)

$$\tilde{f}(k)\langle\psi(k)\rangle = (2\pi)^n f(x)\langle\varphi(-x)\rangle. \tag{1.23}$$

We shall write symbolically $\tilde{f}(k) = \int_{-\infty}^{\infty} f(x) \exp(ikx)\, dx$ whether or not $f(x)$ is summable unless otherwise stated.

This definition results from considering Parseval's relation

$$(2\pi)^{-n} \int_{-\infty}^{\infty} f(x)\varphi(-x)\, dx = \int_{-\infty}^{\infty} \tilde{f}(k)\psi(k)\, dk,$$

valid for absolutely integrable functions.

We note that the Fourier transform $\tilde{f}$ of f in general is an analytic functional, i.e., the left-hand side of (1.23) has the form $\int_{\Gamma} \tilde{f}(k)\psi(k)\, dk$ with some contour Γ in the complex k-plane. The usual rules are valid, e.g., $Ff' = -ikFf$, $(Ff)' = F(ixf)$, $F\sum_0^\infty f_n = \sum_0^\infty Ff_n$.

Let, in particular, $\varphi \in D$, $f \in D'$. Then $\psi \in Z$ and, therefore, $\tilde{f} \in Z'$ is an analytic functional. If $\varphi \in S$, $f \in S'$, then also $\tilde{f} \in S'$ since then $\psi \in S$. Besides the familiar differentiation and multiplication rules, the following results should be noted. If $g(k)$ is regular and is a multiplier in the space $\Psi = F\phi$ consisting of the Fourier transforms of the elements of ϕ, and if $f = F^{-1}g$ (F^{-1} denotes the inverse of F), then for every $h \in \phi'$ we have the convolution rule $F(f * h) = (Ff)\cdot(Fh)$. If f is a multiplier in ϕ, the inverse formula holds, viz., $F(f\cdot h) = (2\pi)^{-n}(Ff) * (Fh)$. For generalizations see §2. If f_λ and $\tilde{g}_\lambda$ depends analytically on the parameter $\lambda \in \Lambda$ and if $\tilde{g}_\lambda = Ff_\lambda$ for $\lambda \in \Lambda_0 \subset \Lambda$, then $\tilde{g}_\lambda = Ff_\lambda$ also for $\lambda \in \Lambda$ by analytic continuation. If $f(x)$ has bounded support or decreases sufficiently rapidly as $|x| \to \infty$, its Fourier transform follows from $f\langle\varphi\rangle$ by replacing $\varphi(x)$ by $\exp ikx$: $\tilde{f}(k) = Ff(x) = f(x)\langle\exp ikx\rangle$.

As a simple example we quote (in R^1)

$$F\delta^{(n)}(x) = (-ik)^n. \tag{1.24}$$

For,

$$(F\delta^{(n)})\langle\psi\rangle = (2\pi)\delta^{(n)}\langle\varphi(-x)\rangle$$
$$= (2\pi)\varphi^{(n)}(0) = \int_{-\infty}^{\infty} (-ik)^n\psi(k)\, dk = (-ik)^n\langle\psi\rangle.$$

Furthermore, by series expansion of exp αx and term-by-term application of the preceding example we obtain

$$(1.25)\qquad Fe^{\alpha x} = 2\pi \sum_0^\infty (-i\alpha)^n \delta^{(n)}(k)/n! = 2\pi\delta(k - i\alpha) \in Z',$$

where α is an arbitrary complex number.

(ix) Let g be a generalized function in R^1. Then the general solution of the equation $x^n \cdot f = g$ is given by

$$f = Pf\,(g/x^n) + f_0,$$

where $Pf\,(g/x^n)$ is a special solution of that equation and $f_0 = \sum_{\mu=0}^{n-1} c_\mu \delta^{(\mu)}(x)$ is the general solution of the homogeneous equation $x^n \cdot f_0 = 0$. For a formal proof just multiply the solution f by x^n and apply (1.10) and (1.11). Similar results hold for analytic functional equations: Let $P(z)$ and $g(z)$ be polynomials in Z', $z = x + iy \in C$. Then the solution of the equation $P(z)f(z) = g(z)$ is given by

$$(1.26)\qquad f(z)\langle\psi\rangle = \left(\int_\Gamma + \int_{\Gamma_0}\right) \frac{g(z)\psi(z)}{P(z)}\,dz,$$

where Γ is a path equivalent to the real axis and Γ_0 a null contour (§1.2). $P(z)f_0(z) = 0$ is solved by $f_0\langle\psi\rangle = \int_{\Gamma_0} (\psi/P)\,dz$. For example, particular solutions of $z^n f = 1$ are the generalized functions

$$(1.27)\qquad f_\pm\langle\psi\rangle = \int_{\Gamma_\pm} z^{-n}\psi(z)\,dz \equiv [z^{-n}]_{\Gamma_\pm}\langle\psi\rangle,$$

where the contours $\Gamma_\pm$ are given in Fig. 1. Then we find

$$(1.28)\qquad \begin{aligned} f_0\langle\psi\rangle &= (f_- - f_+)\langle\psi\rangle = \int_{\Gamma_0} z^{-n}\psi(z)\,dz \\ &= (2\pi i)\,\frac{(-1)^{n-1}}{(n-1)!}\,\delta^{(n-1)}\langle\psi\rangle \end{aligned}$$

as a solution of $z^n f_0 = 0$, with the contour Γ_0 being a circle in the z-plane enclosing the origin (Fig. 1). The solution $f_1 = (f_+ + f_-)/2$ of $z^n f = 1$ equals

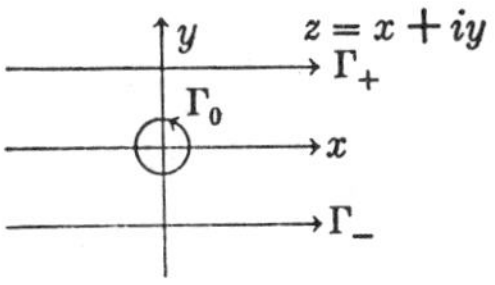

FIG. 1. *The integration contours Γ_+, Γ_- and Γ_0*

the principal value $f_1\langle\psi\rangle = \text{p.v. } x^{-n}\langle\psi(x)\rangle$. Letting $\Gamma_\pm$ in (1.27) approach the real axis we obtain

$$(1.29) \qquad f_\pm(x) = \lim_{y\to 0} f_\pm(z) = \lim_{y\to +0} (x \pm iy)^{-n} = (x \pm i0)^{-n};$$

and combining this with f_0, f_1 it follows that

$$(1.30) \qquad \begin{aligned} f_\pm &\equiv \pm 2\pi i \frac{(-1)^n}{(n-1)!} \delta_\pm^{(n-1)}(x) \\ &= \frac{1}{(x \pm i0)^n} = \text{p.v.} \frac{1}{x^n} \mp \frac{i\pi(-1)^{n-1}}{(n-1)!} \delta^{(n-1)}(x). \end{aligned}$$

(x) The preceding example illustrates the notion of boundary values of analytic functions: $\delta_\pm(x)$ are boundary values of the function $\mp(2\pi i z)^{-1}$ on the real x-axis as $y \to +0$.

The following heuristic argument (which can easily be made rigorous [6]) shows how the concept of boundary values of analytic functions emerges: with $\varphi(x') = \int_{-\infty}^{\infty} \delta(x' - x)\varphi(x)\, dx$ and (cf. (1.30))

$$2\pi i \delta(x' - x) = \lim_{\epsilon\to 0} [(x' - x - i\epsilon)^{-1} - (x' - x + i\epsilon)^{-1}],$$

we obtain

$$(1.31) \qquad \varphi(x') = \lim_{\epsilon\to 0} \frac{1}{2\pi i} \int_{-\infty}^{\infty} \left[\frac{1}{x' - x - i\epsilon} - \frac{1}{x' - x + i\epsilon}\right] \varphi(x)\, dx.$$

Operating on (1.31) with a generalized function $h(x')$ having compact support or decreasing to zero at infinity it follows that

$$(1.32) \quad h(x')\langle\varphi(x')\rangle = \lim_{\epsilon\to 0} \int_{-\infty}^{\infty} \{v(x + i\epsilon) - v(x - i\epsilon)\}\varphi(x)\, dx,$$

where ($z = x + iy$)

$$(1.33) \qquad v(z) = \frac{1}{2\pi i} h(x')\langle(x' - z)^{-1}\rangle$$

is called the indicatrix or generating function of the generalized function $h(x')$, symbolically,

$$(1.34) \qquad v(z) = \frac{1}{2\pi i} \int_{-\infty}^{\infty} \frac{h(x')\, dx'}{x' - z}.$$

Writing (1.32) symbolically in the form

$$(1.35) \qquad h(x) = \lim_{\epsilon\to 0} \{v(x + i\epsilon) - v(x - i\epsilon)\},$$

we see that the generalized function $h(x)$ can be represented in terms of the boundary values of $v(z)$ along the upper and lower boundary of its support (here, the real x-axis). We can generalize these results as follows (see [6]):

Let $f \in K'\{M_p\}$ be a generalized function with support on the real axis and suppose the increase of $f(x)$ be characterized by a function $M, f \approx M$ as $|x| \to \infty$, where M is of the same type as the M_p (or majorizes them). Then there exists an indicatrix $u(z)$, holomorphic outside the support of $f(x)$ and made up of a pair $u = (u_+, u_-)$ of functions $u_+(z)$, $u_-(z)$ holomorphic in the upper and lower half-planes, respectively, such that $f(x)$ is the boundary value of $u(z)$ in the sense that

$$(1.36) \qquad f(x)\langle\psi\rangle = \lim_{\epsilon \to 0} \{u(x + i\epsilon) - u(x - i\epsilon)\}\langle\psi\rangle.$$

$u(z)$ is given by

$$(1.37) \quad u(z) = \begin{Bmatrix} u_+(z), & \operatorname{Im} z > 0 \\ u_-(z), & \operatorname{Im} z < 0 \end{Bmatrix}$$

$$= \frac{M(z)}{2\pi i} f(x')\langle[M(x')(x' - z)]^{-1}\rangle + u_0(z)$$

or, symbolically, by (the principal value p.v. operates on the roots of $M(x')$)

$$(1.38) \qquad u(z) = \frac{M(z)}{2\pi i} \text{ p.v.} \int_{-\infty}^{\infty} \frac{f(x')\,dx'}{M(x')(x' - z)} + u_0(z),$$

where $u_0(z)$ is an arbitrary entire analytic function (which cannot in general be expressed in terms of f). Assuming $\psi \in \phi$ to be analytic ($\phi = Z\{M_p\}$, S_α^β, etc.), (1.36) can be written as an analytic functional,

$$(1.39) \qquad f\langle\psi\rangle = -\int_{\Gamma_0} u(z)\psi(z)\,dz,$$

where Γ_0 is a null contour encircling the support of f. $u_0(z)$ is recognized to be a multiplier in ϕ since $\int_{\Gamma_0} u_0\psi\,dz = 0$.

A simple example is provided by (1.28): the indicatrix

$$v(z) = (-1)^{n+1} n!(2\pi i)^{-1} z^{-n-1}$$

generates $\delta^{(n)}(x)$. Similarly, $u(z) = -(2\pi i)^{-1} \log z$ generates the step function $\theta(x)$.

A theorem, important for many applications, tells that the boundary values $u_\pm(x) = \lim_{y \to +0} u(x \pm iy)$ of $u_\pm(z)$ on the real axis are precisely the (inverse) Fourier transforms $u_\pm = F^{-1} g_\pm$ of generalized functions

$g_\pm = g_\pm(p)$, where $g_{\underset{(-)}{+}}(p) = 0$ for $p \underset{(>)}{<} 0$ and

$$\lim_{|p|\to\infty} |g_\pm(p)| \exp(-\epsilon |p|) = 0$$

for any $\epsilon > 0$: $u_\pm(x) = u(x \pm i0) = \int_{-\infty}^{\infty} g_\pm(p) e^{ipx}\, dp.$

Equation (1.38) is called a spectral representation for $u(z)$, $f(x)$ being the spectral function (a generalized function) and $f(x) = 2i \operatorname{Im} u(x)$ if $u(z)$ is real-analytic (i.e., if $\overline{u(z)} = u(\bar{z})$). Choosing $u_-(z) = 0$ we have $f(x) = u_+(x) = \lim_{\epsilon\to 0} u(x + i\epsilon)$, and inserting this into (1.38) we obtain, by virtue of (1.30),

$$(1.40) \qquad f(x) = \frac{M(x)}{i\pi} \text{ p.v.} \int_{-\infty}^{\infty} \frac{f(x')\, dx'}{M(x')(x' - x)} + u_0(x).$$

Setting $f = \operatorname{Re} f + i \operatorname{Im} f$ and separating real and imaginary parts we obtain the "subtracted" dispersion relations,

$$(1.41) \qquad \operatorname{Re} f(x) = \frac{M(x)}{\pi} \text{ p.v.} \int_{-\infty}^{\infty} \frac{\operatorname{Im} f(x')\, dx'}{M(x')(x' - x)} + \operatorname{Re} u_0(x),$$

$$(1.42) \qquad \operatorname{Im} f(x) = -\frac{M(x)}{\pi} \text{ p.v.} \int_{-\infty}^{\infty} \frac{\operatorname{Re} f(x')\, dx'}{M(x')(x' - x)} + \operatorname{Im} u_0(x),$$

and

$$(1.43) \qquad f(x) = \frac{M(x)}{\pi} \text{ p.v.} \int_{-\infty}^{\infty} \frac{\operatorname{Im} f(x')\, dx'}{M(x')(x' - x - i0)} + u_0(x).$$

For example, if $f(x) \approx x^{n-\epsilon} \in S'$ as $|x| \to \infty$ ($\epsilon > 0$), we may choose $M = (x - x_0)^n$ (x_0 real) obtaining

$$(1.44) \qquad \begin{aligned} f(x) = \frac{(x - x_0)^n}{\pi} \underset{(x'=x_0)}{\text{p.v.}} \int_{-\infty}^{\infty} & \frac{\operatorname{Im} f(x')\, dx'}{(x' - x_0)^n (x' - x - i0)} \\ & + \sum_{\mu=0}^{n-1} \frac{f^{(\mu)}(x_0)(x - x_0)^\mu}{\mu!}. \end{aligned}$$

These relations furnish a substitute for the Hilbert transform $\int_{-\infty}^{\infty} [g(x')/(x' - x)]\, dx'$ in case of increasing spectral functions $g(x')$, containing a number of unknown "subtraction" constants $(f^{(\mu)}(x_0))$ which cannot be expressed in terms of f and must be determined from experiment in physical situations. If, as in (1.44), $u_+ \in S'$, then one also obtains (1.44) by applying Cauchy's theorem to $u(z + i\epsilon')/M(z + i\epsilon')$ in the upper half-plane $\operatorname{Im} z > 0$ with the path of integration consisting of a straight line above the real axis and closed by a semicircle with a radius tending to infinity, letting finally $\epsilon' \to +0$ and utilizing (1.30) (see [6]).

There are many other important concepts in generalized function theory (such as regular and singular variable transforms) having immediate applications but we shall defer the discussion. The only important notion to be introduced yet here is that of positive and positive-definite generalized functions.

(xi) We shall call a generalized function f positive, and write $f \geqq 0$, if $f\langle\varphi\rangle \geqq 0$ for every positive φ. A generalized function $f \in D'$ is called positive-definite if $f\langle\varphi * \hat{\varphi}\rangle \geqq 0$ for every $\varphi \in D$ where $\hat{\varphi}(x) = \overline{\varphi(-x)}$ and $\varphi * \hat{\varphi}$ denotes the convolution $\varphi * \hat{\varphi} = \int_{-\infty}^{\infty} \varphi(y)\overline{\varphi(y - x)}\, dy$. Symbolically we shall write $f\langle\varphi * \hat{\varphi}\rangle \geqq 0$ in the form

$$\iint_{-\infty}^{\infty} f(x - y)\varphi(x)\bar{\varphi}(y)\, dx\, dy \geqq 0. \tag{1.45}$$

If $f \in Z'$, the positive-definiteness tells that $f\langle\psi * \hat{\psi}\rangle \geqq 0$, where $\hat{\psi}(z) = \overline{\psi(-\bar{z})}$. The importance of these notions lies in Bochner's theorem and in its generalizations.

THEOREM. (a) *The Fourier transform of any positive measure (or of any positive generalized function of D') is a positive-definite generalized function of Z', and conversely.* (b) *The Fourier transform of any positive tempered measure is a positive-definite generalized function of D', and conversely.*

The formal proof of this theorem follows by simply inserting the Fourier representation of f into the positive-definiteness condition.

2. Recent developments in generalized function theory. In this section we shall discuss some new developments in the theory of generalized functions. Most of them originated from physical considerations which we shall briefly discuss in the next sections. There are many open problems in this and the following sections which we recommend to the attention of mathematicians.

2.1. Integral representation of singular generalized functions. If in the identity (1.19) one replaces λ by $\lambda - z$, multiplies both sides by a^z/z and finally takes on both sides of the resulting identity the residue at $z = 0$ (z being a complex variable), one obtains for x_+^λ and similarly for x_-^λ the integral representation (valid for all complex λ, cf. below)

$$x_\pm^\lambda\langle\varphi\rangle = \operatorname*{Res}_{z=0}\left[\frac{1}{z}\int_{-\infty}^{\infty} x_\pm^\lambda (a/x)^z \varphi(x)\, dx\right]. \tag{2.1}$$

The left-hand side actually is $Pf\, x_\pm^\lambda$ for $a = 1$ but according to our convention we shall drop the Pf-symbol. $x_\pm^\lambda$ in the integrand on the right-hand side is the classical function. Equation (2.1) has to be interpreted as follows: one first has to compute the x-integral $F(z) = z^{-1} \int_{-\infty}^{\infty} x_{\pm}^{\lambda}(a/x)^{z}\varphi(x)\,dx$ for Re z sufficiently large and negative (where it converges also for Re $\lambda < 0$), then to continue the resulting function $F(z)$ analytically into a neighborhood of the origin $z = 0$ and finally to apply the residue. The result obviously coincides with the analytic continuation of $x_{\pm}^{\lambda}\langle\varphi\rangle$ in λ from Re $\lambda > 0$ into the complex λ-plane except for the points $\lambda = -1, -2, \cdots$, where $F(z)$ has simple poles. Furthermore, the left-hand side of (2.1) is independent of the arbitrary positive constant a unless $\lambda = -1, -2, \cdots$. Contrary to (1.16), (2.1) also makes sense for $\lambda = -1, -2, \cdots$, but in that case (and only then) does the left-hand side depend on the arbitrary scaling parameter a. If we look more closely, this dependence on a takes place if and only if the classical integral $\int_{-\infty}^{\infty} x_{\pm}^{\lambda}\varphi(x)\,dx$ exhibits logarithmic divergences (i.e., for $\lambda = -1, -2, \cdots$). If we choose $a = 1$, we obtain exactly the usual representation for $Pf\, x_{\pm}^{-n}$, viz., (1.18). However, a glance at the Fourier transform of $Pf\, x_{\pm}^{-n}$ as defined by (1.18) (and obtained by replacing $\varphi(x)$ in that equation by exp ipx) shows that the definition (1.18) is rather unphysical. For, in physics, the variable x has dimension of length and the variable p in $Fx_{\pm}^{-n} \approx p^{n-1} \log (p \pm i0)$ has the dimension of a reciprocal length, whereas the argument of any transcendental function, in particular, that of the logarithm, must be dimensionless. Consequently, another length, a, must somehow be introduced in the sense that $\log p \to \log ap$ receives a dimensionless argument. Precisely this is guaranteed by our preceding procedure.

One may look at this also in another way: the generalized function $Pf\,(x_{\pm}^{\lambda})$ is not homogeneous if $\lambda = -1, -2, \cdots$ and if one defines it by (1.18). For by a change of scale $x \to \alpha x$ we have ($\alpha > 0$)

$$Pf\,(\alpha x)_{\pm}^{-n} = \alpha^{-n} Pf\,(x_{\pm}^{-n}) \pm \alpha^{-n} \frac{(-1)^{n-1}}{(n-1)!} \log \alpha \cdot \delta^{(n-1)}(x).$$

In other words, $Pf\, x_{\pm}^{-n}$ depends on the coordinate unit chosen. Let $f(x)$ be an ordinary function singular at $x = 0$. A "regularization" of $f(x)$ is, by definition, a functional f_R which coincides with $f(x)$ everywhere except at $x = 0$, i.e., $f_R\langle\varphi\rangle = f\langle\varphi\rangle$ for all those $\varphi(x)$ which vanish at $x = 0$ together with a sufficiently large number of derivatives. $Pf\, x_{+}^{-n}$, (1.18), is such a regularization and any regularization is obtained by adding to $Pf\, x_{+}^{-n}$ a term $\sum_{0}^{n-1} c_{\mu}\delta^{(\mu)}(x)$ with arbitrary coefficients c_{μ}. Regularizations are in fact constructive realizations of the Hahn-Banach extension theorem. In this "regularization philosophy" we may say, therefore, that the extension Ef of $f(x)$, $x \neq 0$, into a linear and continuous functional given by $Pf\, x_{\pm}^{-n}$ does not commute with the dilatation $\mathbf{U}_{\alpha}$, $\mathbf{EU}_{\alpha}f \neq \mathbf{U}_{\alpha}\mathbf{E}f$, where $\mathbf{U}_{\alpha}f\langle\varphi\rangle$

$= f(\alpha x)\langle\varphi\rangle = \alpha^{-1}f\langle\varphi(x/\alpha)\rangle$. It is then appropriate to consider a class of extensions $\mathbf{E}_\alpha$ defined by $\mathbf{E}_\alpha f = \mathbf{U}_\alpha^{-1}\mathbf{E}\mathbf{U}_\alpha f$. This extension, depending on a parameter α (a "unit of length"), yields, if applied to the function x_+^{-n}, also the representation (2.1) (see [9]).

Once the role of the scaling parameter a (of dimension of length if x is a length) is clearly recognized, it is obvious that $x_\pm^{-n} = x_\pm^{-n}|_a$, defined by (2.1) is homogeneous in the sense that a change of scale $x \to \alpha x$ accompanied by a corresponding change in a, $a \to \alpha a$, which is physically necessary, results in the formula

$$(2.2) \qquad (\alpha x)_\pm^{-n}|_{\alpha a} = \alpha^{-n}x_\pm^{-n}|_a .$$

We may sum these results as follows.

Statement. Let $f(x)$ be a function of the single variable x having an algebraic singularity at $x = 0$. Then the generalized function f is given by the integral

$$(2.3) \qquad f\langle\varphi\rangle = \operatorname*{Res}_{z=0}\left[\frac{1}{z}\int_{-\infty}^{\infty} f(x)\left(\frac{a}{x}\right)^z \varphi(x)\,dx\right],$$

where a is an arbitrary scaling parameter (of dimension of length if x is a length) on which the left-hand side depends, if and only if the classical integral $\int_{-\infty}^{\infty} f(x)\varphi(x)\,dx$ exhibits logarithmic divergences. In (2.3) one first has to evaluate the x-integral for Re z sufficiently large and negative, where it converges, then to continue the resulting function of z into a neighborhood of the origin $z = 0$ and finally to take the residue. In case of functions of the type $x_\pm^{-n}\log^\mu x_\pm$, etc., (2.3) yields a generalization of the conventional Pf-notion. If $f(x)$ has an essential singularity at $x = 0$, (2.3) is to be applied to every term in the expansion of f.

If $g(r)$, $r = |\mathbf{x}|$, $x \in R^n$, has an algebraic singularity at $r = 0$, then

$$(2.4) \qquad g\langle\varphi\rangle = \operatorname*{Res}_{z=0}\left[\frac{1}{z}\int_{-\infty}^{\infty} g(r)\left(\frac{a}{r}\right)^z \varphi(x)\,d^nx\right].$$

If $f(x)$ and $g(r)$ decrease at infinity sufficiently rapidly, their Fourier transforms follow from (2.3) and (2.4) by replacing $\varphi(x)$ by exp ikx. In particular ($k = p + iq$),

$$(2.5) \qquad \tilde{f}(k) = Ff(x) = \operatorname*{Res}_{z=0}\left[\frac{1}{z}\int_{-\infty}^{\infty} f(x)\left(\frac{a}{x}\right)^z e^{ikx}\,dx\right],$$

while for Fg we obtain the generalized Bochner formula

$$(2.6) \quad \tilde{g}(p) = Fg(r)|_{k=p} = \operatorname*{Res}_{z=0}\left[\frac{(2\pi)^{n/2}}{zp^{n/2-1}}\int_0^\infty r^{n/2}f(r)\left(\frac{a}{r}\right)^z J_{n/2-1}(pr)\,dr\right].$$

For example, by straightforward calculation we find

$$(2.7)\quad Fx_{\pm}^{\lambda}|_{k=p} = \begin{cases} \pm ie^{\pm i\lambda\pi/2}\Gamma(\lambda+1)(p \pm i0)^{-\lambda-1}, & \lambda \neq -1, -2, \cdots, \\ \dfrac{(\pm i)^{n-1}p^{n-1}}{(n-1)!}[\log|ap| \mp i\pi/2 - \Gamma'(n)/\Gamma(n)], & \\ & \lambda = -n = -1, -2, \cdots. \end{cases}$$

2.2. Products of generalized functions. Quantum field theory suffers strongly from undefined products of generalized functions such as $\theta(x)\cdot x^{-n}$, $\delta(x)\cdot x^{-n}$, etc. A general product theory for distributions has been developed by König [12] and the author [11]. It suffers from arbitrariness as do the subtracted dispersion relations in §1.3. It is the aim of this and the next section to get rid of this arbitrariness.

Let us first sketch the method of [11] and [12] for defining products of generalized functions based on the Hahn-Banach extension theorem. Let $g, f \in D'$ and assume that neither g nor f is a multiplier in D so that the rule (1.10) fails to hold. For example, let $g = \theta$ and $f = \delta$. If θ were a multiplier in D we could apply (1.10) to obtain ($\psi \in D$, $\circ$ denotes the product)

$$(2.8)\qquad h\langle\psi\rangle \equiv (\theta\circ\delta)\langle\psi\rangle = \delta\langle\theta\psi\rangle = \theta(0)\psi(0) = \theta(0)\delta\langle\psi\rangle.$$

This is a meaningless result since $\theta(0)$ is not defined. Equation (2.8) becomes meaningful, however, for all $\psi(x) \in D$ with $\psi(0) = 0$, say, $\psi \in U \subset D$. Then $h\langle\psi\rangle = 0$ on U. Assume now that the product $h = \theta\circ\delta$ is a linear continuous functional in D' (one can indeed show that the algebra of distributions can be mapped into D') known on $U \subset D$. Then $h \in D'$ is known on the whole space D if its value is known on any $\varphi_0 \in D$, $\varphi_0 \notin U$, i.e., if h is given an arbitrary, value on φ_0. Since $\varphi(x) = \{\varphi(0)\varphi_0(x) + \psi(x)\} \in D$ with $\varphi_0(0) = 1$, we obtain $h\langle\varphi\rangle = \varphi(0)h\langle\varphi_0\rangle + h\langle\psi\rangle$, and since $h\langle\varphi_0\rangle$ was arbitrary, say, $h\langle\varphi_0\rangle = c$, it follows that $h \equiv \theta\circ\delta = c\delta + 0 = c\delta$, c being an arbitrary constant. Similarly, one finds that $\delta(x)\circ x^{-1} = -\delta'(x) + c\delta(x)$. One can make a rigorous product theory that way. Let $f, g \in D'$ and let $U = U[f, g] \subset D$ be a subspace of D, depending on f and g, defined as follows: $\psi(x)$ is an element of U if for every x there exists a nonnegative integer n such that either (i) $f \in D_n'$ is a distribution of order n and $\psi\cdot g \in E_n$ is an n-times continuously differentiable function, or (ii) $f \in E_n$ and $\psi\cdot g \in D_n'$. Then, for every x there exists either (i) the functional $f\langle\psi\cdot g\rangle$, or (ii) the functional $(\psi\cdot g)\langle f\rangle$. We define a functional $h\langle\psi\rangle$ for each x according to $h\langle\psi\rangle = f\langle\psi\cdot g\rangle$ in case (i), and $h\langle\psi\rangle = (\psi\cdot g)\langle f\rangle$ in case (ii). Then the product $f\circ g$ is defined on U by the formula $(f\circ g)\langle\psi\rangle \triangleq h\langle\psi\rangle$. From the requirement (i) or (ii) certain conditions on $\psi \in U$ emerge relating to the vanishing of $\psi(x)$ in the singular points of f and g. By means of the formula

$$\varphi(x) = \sum_{n=0}^{2m} \varphi^{(n)}(0)\varphi_n(x) + \psi(x),$$

$$\varphi \in D, \quad \psi \in U, \quad \varphi_n \in D, \quad \varphi_n \notin U, \quad \varphi_n^{(\mu)}(0) = \delta_{n\mu},$$

the product can then be extended to the whole space D. The result is of the type $f \circ g = h + \sum a_{\nu\mu}\delta^{(\nu)}(x - x_\mu)$ with arbitrary coefficients $a_{\nu\mu}$, x_μ being the common singular points of f and g. If g is a multiplier, then we have $g \circ f = g \cdot f$. One can then show this: Let $a(x)$ be a multiplier in D; then the multiplication rule $a \circ (f \circ g) = f \circ (a \cdot g)$ holds. This enables us to construct the product $h = f \circ g$ in terms of the division problem (cf. §1.3) $a \cdot h = f \cdot (a \cdot g)$: To find h one forms $f \circ (a \cdot g)$ with an arbitrary multiplier $a(x)$, requires that $a(x)$ be such that $f \circ (ag) = f \cdot (a \cdot g)$ exists and solves $a \cdot h = f \cdot (a \cdot g)$ with this a for h. The resulting products in general are neither commutative nor associative nor do they satisfy the conventional classical rules for variable transformations, e.g., even if f, g are homogeneous generalized functions, the product $f \circ g$ may be nonhomogeneous unless the $a_{\mu\nu}$ are put equal to zero or transform as densities.

Products of distributions likewise may be defined in terms of products of the indicatrices (§1.3) which generate the factors of the product. Furthermore, if $\tilde{f} = Ff$, $\tilde{g} = Fg$, where both $\tilde{f}$, $\tilde{g}$ have their supports bounded to the left or right and form a convolution algebra in p-space, the product $f \circ g$ may be defined by the right-hand side of the formula (cf. (viii) of §1.3) $f \circ g = (2\pi)^{-n}F^{-1}\{(Ff) * (Fg)\}$. For example, with $f = \delta_\pm^{(n)}$, $g = \delta_\pm^{(m)}$, (1.30), one finds that $F\delta_\pm^{(n)}(x) = (-ip)^n\theta(\pm p)$, whence

$$\delta_\pm^{(n)}(x) \circ \delta_\pm^{(m)}(x) = \pm\frac{n!m!}{2\pi i(n + m + 1)!}\,\delta_\pm^{(n+m+1)}(x)$$

without encountering any arbitrariness.

A general "subtraction-free" product can be defined for the physically important class of generalized functions, which are, except for isolated points, boundary values of analytic functions and step functions, by the formula

$$(2.9) \qquad (f(x) \circ g(x))\langle\varphi\rangle = \operatorname*{Res}_{z=0}\left[\frac{1}{z}\int_{-\infty}^{\infty} f(x)g(x)\left(\frac{a}{x}\right)^z \varphi(x)\,dx\right]$$

with an interpretation as in (2.3). The most general extension differs from (2.9) by a derivative polynomial of the type $\sum a_{\mu\nu}\delta^{(\nu)}(x - x_\mu)$. The formula (2.9) represents essentially the result of a definition of products by analytic continuation. Apart from products of propagators such as $\delta_\pm^{(n)} \circ \delta_\pm^{(m)}$—which follows from (2.9) too (with $f = (x \pm i0)^{-n}$, $g = (x \pm i0)^{-m}$)—the product of a generalized function f with the step function θ is physically most important (§3, §4) since essentially all divergences of field theory originate from letting this product be undefined. From (2.9) we obtain now immediately

$$(2.10)\qquad (\theta(x)\circ f(x))\langle\varphi\rangle = \operatorname*{Res}_{z=0}\left[\frac{1}{z}\int_0^\infty f(x)\left(\frac{a}{x}\right)^z \varphi(x)\,dx\right],$$

that is to say,

$$(2.11)\qquad \theta(\pm x)\circ f(x) = \operatorname*{Res}_{z=0}\left[\frac{1}{z}\theta(x)f(x)\left(\frac{a}{x}\right)^z\right] = f_\pm(x).$$

In other words, the product $\theta(\pm x)\circ|x|^\lambda$ equals $x_\pm^\lambda$ for all λ and is given by continuing analytically in λ the well-defined product $\theta(x)\cdot|x|^\lambda$ from $\operatorname{Re}\lambda > 0$ to the λ-plane with the appropriate definition at the poles $\lambda = -1, -2, \cdots$ provided by the residue formula. This product definition is equivalent to what one obtains by differentiating the regular functionals $x_\pm^\lambda$ and $\log x_\pm$ ($\lambda > 0$) according to §1.3. The only arbitrariness involved is that of the scaling parameter a which, as we have seen in §2.1, is a physically necessary quantity.

2.3. Generalized Hilbert transforms and unsubtracted dispersion relations. In (x) of §1.3 we have found a "subtracted" substitute for the nonexistent Hilbert transform $\int_{-\infty}^{\infty} g(x')(x'-x)^{-1}\,dx'$ in the case of a spectral function $g(x')$ increasing as $|x'|\to\infty$. By starting from the convolution product

$$(2.12)\qquad x^{-1} * x^\alpha = \int_{-\infty}^{\infty}\frac{x'^\alpha\,dx'}{x-x'}$$

defined for $\operatorname{Re}\alpha < 0$, we can define the Hilbert transform of x^α, $\alpha > 0$, by analytically continuing (2.12) in α to the complex α-plane except for the points $\alpha = 0, 1, \cdots$. At $\alpha = 0, 1, 2, \cdots$ the pole-free, i.e., constant, part in (2.12) is projected out by means of our residue procedure, and we arrive at the defining formula for the Hilbert (and similarly for the Stieltjes) transform of a generalized function $f(x)$ increasing at infinity, viz.,

$$(2.13)\qquad \frac{1}{x} * f(x) = \operatorname*{p.v.}_{(x'=x)}\int_{-\infty}^{\infty}\frac{f(x')\,dx'}{x-x'} \triangleq \operatorname*{Res}_{z=0}\left[\frac{1}{z}\int_{-\infty}^{\infty}\frac{f(x')\,(ax')^z dx'}{x-x'}\right],$$

(symbolic integral)

where a is an arbitrary scaling parameter (of dimension of length since in formulas of this type x in general has dimension of momentum, i.e., of a reciprocal length). Equation (2.13) is evaluated by carrying out the integral for Re z sufficiently large and negative, continuing the resulting function of z to $z = 0$ and taking the residue. The left-hand side depends on a if and only if the classical formal integral $\int_{-\infty}^{\infty} f(x')(x - x')^{-1}\,dx'$ exhibits logarithmic divergences.

The following examples are easily verified:

$$\int_0^{\infty} \frac{x'^{\lambda}\,dx'}{x' - x \pm i0} = \operatorname*{Res}_{z=0}\,[\Gamma(-\lambda - z)\Gamma(\lambda + z + 1) \cdot (-x \mp i0)^{\lambda+z} a^z/z], \tag{2.14}$$

$$f(x) = \operatorname*{Res}_{z=0}\left[\frac{1}{\pi z}\int_{-\infty}^{\infty} \frac{\operatorname{Im} f(x')(ax')^z\,dx'}{x' - x - i0}\right], \qquad f \in S'. \tag{2.15}$$

Comparing the above results with those of (x) of §1.3 one sees that the "subtraction" constants have been avoided. The connection of the generalized Hilbert transform with the product theory of §2.2 is easily established: Consider the product $f_+(x) = \theta(x)\cdot f(x)$ defining a generalized function f_+ having its support on the right (left) semiaxis and assume that f is regular. The Fourier transform $\tilde{f}_+ = Ff_+$ represents, with k replaced by z, the indicatrix $u_+(z)$ of §1.3, and we have $\tilde{f}_+(p) = \delta_-(p) * \tilde{f}(p)$ in virtue of $F\theta(x) = \delta_-(p)$, $Ff = \tilde{f}$. Therefore,

$$\tilde{f}_+(p) = \frac{1}{2\pi i}\int_{-\infty}^{\infty} \frac{\tilde{f}(p')\,dp'}{p - p' - i0} \tag{2.16}$$

exists classically whenever the product $f_+ = \theta \cdot f$ exists in the sense of (1.10). In most of the physically interesting systems, $f(x)$ is singular at $x = 0$ and $\tilde{f}(p) = Ff(x)$ increases as $|p| \to \infty$ so that the nonexistence of the product $f_+(x) = \theta(x)\cdot f(x)$ in the sense of (1.10) implies the nonexistence of the Hilbert transform (2.16), unless one adopts the indicatrix formalism (with its arbitrariness) which implies directly the general product theory of §2.2 if we remember that, e.g., with $f = x^{-n}$, $\theta(x) \circ x^{-n} = Pf\ x_+^{-n} + \sum_0^{n-1} c_\mu \delta^{(\mu)}(x)$ with arbitrary constants c_μ. Things become different and free from the arbitrariness if the Hilbert transform (2.16) is defined by (2.13), say, by analytic continuation, and the product $\theta(x) \circ f(x)$ corresponding to this $\tilde{f}_+(p)$ is then given by formula (2.10) without any arbitrariness, except for the scaling parameter a.

Whenever we speak about unsubtracted dispersion relations in what follows we mean by this term relations defined by the preceding continuation or residue procedure. The relation (2.13) holds, first of all, for tempered distributions, e.g., for functions polynomially bounded at infinity. For nontempered distributions such as $f(x) = \sum_0^{\infty} c_n x^{\eta_n} \log^{\zeta_n} |x| (\eta_n \to \infty$ as $x \to \infty$, $\zeta_n = 0, 1, \cdots)$ the residue operation has to be applied, by definition, term by term in the expansion of $f(x)$. If, e.g., $f = \exp |x|^\alpha$, $\alpha > 0$, a different definition may be applied, viz.,

$$x^{-1} * \exp(|x|^\alpha) = \operatorname*{Res}_{z=0}\left[\frac{a^z}{z}\int_{-\infty}^{\infty}\frac{\exp[(1-z)|x'|^\alpha]\,dx'}{x-x'}\right],$$

but the physical examples known thus far (§3, §4 and [7], [14]) indicate some sort of superiority of the term-by-term application of (2.13) to the expansion of $\exp |x|^\alpha$. The problem, of course, deserves further investigation.

2.4. Infinite series of derivatives of delta functions and high frequency/energy bounds. Let $f(x)$ determine a functional $f \in D'$, and let its Taylor series $f(x) = \sum_0^\infty a_n x^n$ converge to $f(x)$, $x \in R^1$, assuming that $f(x)$ is an entire analytic function of $z = x + iy$. Applying (1.24) to the individual terms (noting that the Fourier operator F commutes with summation in the sense of generalized functions) we obtain for $g = Ff$ the series

$$g(k) = Ff(x) = 2\pi \sum_{n=0}^{\infty} (-i)^n a_n \delta^{(n)}(k), \tag{2.17}$$

i.e.,

$$g\langle\psi\rangle = 2\pi \sum_{n=0}^{\infty} i^n a_n \psi^{(n)}(0), \qquad \psi \in Z\{M_p\}. \tag{2.18}$$

By Cauchy's integral formula, applied to $\psi^{(n)}(k)$, the expression (2.18) may be put into the form

$$g\langle\psi\rangle = \int_{\Gamma_0} G(k)\psi(k)\,dk, \tag{2.19}$$

where Γ_0 is a closed contour (a null contour, §1.2) which lies entirely in the region of convergence of the series

$$G(k) = \sum_{n=0}^{\infty} i^{n-1} a_n n!\, k^{-n-1}, \qquad |k| > R^{-1}, \tag{2.20}$$

under the integral sign in (2.19). Suppose the radius of convergence, $R = [\limsup_{n\to\infty} \sqrt[n]{|a_n|n!}]^{-1}$, of this series to be different from zero. Then the contour Γ_0 may be taken to be any which encloses the singular points of $G(k)$. Defining now the support of the analytic functional g to be the smallest closed region in which g is concentrated [6], the support of g is the smallest closed region to which the contour can be shrunk (i.e., the complement of the region of analyticity of the inverse Borel transform G of f). If $R = 0$, g has unbounded support, in which case the representation (2.18) must be analyzed in terms of generalized moments [6].

Using Cauchy's convergence test one sees that $G(k)$ converges for $|k| > 0$, and the support of $g(k)$ is the origin $k = 0$ if

$$\limsup_{n\to\infty} \sqrt[n]{|a_n|\, n!} = 0, \tag{2.21}$$

i.e., if $f(x)$ has order of growth $\rho < 1$. For the case $\rho = 1$ (generalized Paley-Wiener theorem) and $\rho > 1$, see [6].

An example is

$$f(x) = \sum_{n=0}^{\infty} \lambda^n x^n / n!(n+1)!, \qquad g\langle\psi\rangle = -\frac{1}{\lambda}\int_{\Gamma_0} dk\, e^{i\lambda/k}\psi(k) \tag{2.22}$$

with any circle Γ_0 enclosing the origin. Quite generally, $G(k)$ has an essential singularity at $k = 0$ if the order of growth ρ of f is less than one as $|x| \to \infty$. From (2.21) we see that in the "local case", where g is localized at the origin (as are the individual terms $\delta^{(n)}$), the Fourier transform $f = F^{-1}g$ of g is bounded for $|x| \to \infty$ by ($\epsilon > 0$)

$$\lim_{|x|\to\infty} |f(x)| \exp(-\epsilon |x|) = 0. \tag{2.23}$$

This is the high frequency/energy bound which plays a fundamental role in passive system theory and quantum field theory (§3, §4).

2.5. Formal series and all that. Formal series, rigged Hilbert spaces and generalized eigenfunctions promise to become important tools in theoretical physics. A few words on these entities may suffice here: The eigenfunctions h_n of the eigenvalue problem $Hh_n = E_n h_n$ of the linear harmonic oscillator in quantum mechanics are given by the Hermite functions $h_n(x)$. With these one may construct formal series with constant coefficients a_n, viz., $f \sim \sum_0^\infty a_n h_n(x)$, and these series (or the one-column matrix (a_n)) may be viewed as representing generalized functions which ultimately amounts to an algebraization of analysis. If the a_n are rapidly decreasing, the set of $f \sim \sum_0^\infty a_n h_n$ is isomorphic to the space S, the set of formal series with $\sum_0^\infty |a_n|^2 < \infty$ is isomorphic to $H = L^2$, and the set of $\sum_0^\infty a_n h_n$ with slowly increasing a_n is isomorphic to the space S' of tempered distributions once a certain natural topology is introduced in the space of formal series. The triplet $S \subset H \subset S'$ forms a rigged Hilbert space of which, e.g., the delta function $\delta(x - x_0)$ as an eigenfunction of x is a legitimate element. Generalized (infinite) numbers, commutative nuclear Lie groups and other entities can conveniently be discussed in these terms [6], [15], [16].

3. Passive system theory. Suppose an input signal $i(t)$ is fed into a physical system (a "black box"), which responds to this stimulus by producing an output $o(t)$, t being the time. If the system is nonlinear, then $o(t)$

is related to $i(t)$ by the formula

$$(3.1)\quad o(t) = \int_{-\infty}^{\infty} S_1(t, t_1)i(t_1)\,dt_1 + \iint_{-\infty}^{\infty} S_2(t, t_1, t_2)i(t_1)i(t_2)\,dt_1\,dt_2 + \cdots,$$

where the generalized functions $S_n(t, t_1, \cdots)$ represent the physical system. Expanding $i(t)$ and $o(t)$ in terms of orthonormal oscillator functions $h_n(t)$, $i(t) = \sum_0^\infty a_n h_n(t)$ and $o(t) = \sum_0^\infty b_n h_n(t)$, and assuming that the b_n are Taylor expansions of the a_n, $b_n(a_0, a_1, \cdots) = b_n(0, 0, \cdots) + \sum c_n^{\,i} a_i + \sum c_n^{\,ik} a_i a_k + \cdots$, the S_n can be expressed in terms of the h_n according to $S_n(t, t_1, \cdots, t_n) = \sum_{i,k,\ldots} c_n^{\,ik\cdots} h_n(t)h_i(t_1)h_k(t_2) \cdots$. If one subjects the system to the postulates of (i) time-invariance (i.e., if $i(t)$ produces $o(t)$, then $i(t-\tau)$ produces $o(t-\tau)$), (ii) causality (i.e., if $i(t) = 0$ for $t < 0$, then $o(t) = 0$ for $t < 0$), and (iii) passivity (or unitarity) (i.e., the integrated output intensity is less than or equal to the integrated input intensity), viz.,

$$(3.2)\qquad \int_{-\infty}^{\infty} [|\,i(t)\,|^2 - |\,o(t)\,|^2]\,dt \geqq 0,$$

then one can develop a general structure analysis of the system under investigation. However, the result is an empty scheme which has to be filled by specific dynamic assumptions.

Things become particularly simple in the case of a linear system characterized by $S_1 = S$, $S_2 = S_3 = \cdots = 0$. Then by [4], [5] and the references therein, the axioms (i) and (ii) imply that $S(t, t_1) = S(t - t_1)$, $S(t) = 0$ for $t < 0$: $o(t) = S(t) * i(t)$, symbolically,

$$(3.3)\qquad o(t) = \int_{-\infty}^{\infty} S(t - t')i(t')\,dt'$$

with

$$(3.4)\qquad S(t) = 0 \quad \text{for} \quad t < 0:\ S(t) = \theta(t)\cdot s(t).$$

The system function (or "S-matrix") $S(t)$ is assumed to be a generalized function. $o(t)$, $i(t)$ usually are assumed to be in L^2 but may be viewed as generalized functions too if the products $|\,o\,|^2$, $|\,i\,|^2$, etc., are interpreted in the sense of §2.2. With

$$f\langle\bar{g}\rangle = \int f\bar{g}\,dt = \delta(t)\langle f^+ * \bar{g}\rangle, \qquad f^+(t) = f(-t), \qquad \hat{f}(t) = \overline{f(-t)}$$

and considering $i(t)$ as a test function, $i^+(t) = \varphi(t)$, the passivity or unitarity postulate (iii) yields for S the positive-definiteness condition (§1.3) $(\delta - S * \hat{S})\langle\varphi * \hat{\varphi}\rangle \geqq 0$, and Bochner's theorem (see (xi) of §1.3) says then that ($\tilde{S}(k) = FS(t)$)

$$\tilde{S}(k)\overline{\tilde{S}(\bar{k})} \leqq 1. \tag{3.5}$$

If in (3.2) one replaces the upper boundary $+\infty$ by t, one finds that this new passivity condition implies the causality condition (ii). $\tilde{S}(k)\overline{\tilde{S}(\bar{k})} = 1$ means that the system is lossless or satisfies elastic unitarity.

Introducing the variables $j(t) = i(t) - o(t)$ ("current") and $v(t) = i(t) + o(t)$ ("voltage") and the scattering amplitude ("augmented admittance") $T(t)$, $S(t) = \delta(t) + 2iT(t)$, (3.5) implies that for real $k = p$ ($k = p + iq$), $\operatorname{Im} \tilde{T}(p) \geqq |\tilde{T}(p)|^2 > 0$. Assuming further that $S(t), j(t), v(t)$ form a convolution algebra we can write $v(t) = Z(t) * j(t)$, $j(t) = Y(t) * v(t)$, where $Z = (-iT)^{-1} - \delta$ is the impedance and $Y = -i(\delta + iT)^{-1} * T$ the admittance. In what follows we shall further assume that $S(t)$ is real. The passivity condition (3.2) yields for $Z(t)$ the positive-definiteness relation

$$\operatorname{Re} \int_{-\infty}^{\infty} Z(t - t')j(t)\overline{j(t')}\, dt\, dt' \geqq 0. \tag{3.6}$$

3.1. Dispersion theory of passive systems. From the causality condition (ii), i.e., (3.4), it follows in virtue of the theorem mentioned in (x) of §1.3 that

$$\tilde{S}(k) = \int_0^{\infty} S(t) \exp ikt\, dt \tag{3.7}$$

is analytic in the half-plane $\operatorname{Im} k > 0$, $k = p + iq$, and is real on the imaginary axis since $S(t)$ was assumed to be real. Obviously, $\tilde{S}(p)$ is the boundary value of $\tilde{S}(k)$, $\tilde{S}(p) = \lim_{q\to+0} \tilde{S}(p + iq)$. Thus, in the sense of (x) in §1.3, $\tilde{S}(k) \leftrightarrow u_+(z)$ and $\tilde{S}(p) \leftrightarrow u_+(x)$. Therefore, the axioms listed above imply for $\tilde{S}(p)$ the dispersion relations (1.41), (1.42) and the spectral representation for increasing $\tilde{S}$,

$$\tilde{S}(p) = \frac{M(p)}{\pi} \text{ p.v.} \int_{-\infty}^{\infty} \frac{\operatorname{Im} \tilde{S}(p')\, dp'}{M(p')(p' - p - i0)} + u_0(p), \tag{3.8}$$

where $M(p)$ is any weight function such that $[\operatorname{Im} \tilde{S}(p)/M(p)] \to 0$ as $|p| \to \infty$ and $u_0(p)$ is an arbitrary entire function (a polynomial if $\tilde{S}$ is tempered). From the reality of $S(t)$ it follows that $\operatorname{Im} \tilde{S}(-p) = -\operatorname{Im} \tilde{S}(p)$, whence $\operatorname{Im} \tilde{S}(p) = \pi\epsilon(p)\rho(p^2)$ with $\epsilon(p) = \operatorname{sgn} p$ and $\rho(p^2)$ being a spectral function. Inserting the identity

$$\operatorname{Im} \tilde{S}(p) = \pi\epsilon(p)\rho(p^2) = \pi\epsilon(p) \int_0^{\infty} \rho(\omega^2)\delta(p^2 - \omega^2)\, d\omega^2 \tag{3.9}$$

into (3.8) we obtain, with $M(p) = \tilde{M}(p^2)$,

$$(3.10) \qquad \tilde{S}(p) = \tilde{M}(p^2) \int_0^\infty \frac{\rho(\omega^2)\, d\omega^2}{M(\omega^2)[\omega^2 - p^2 - ip0]} + u_0(p).$$

Since $\tilde{S}_0(p, \omega^2) = [\omega^2 - p^2 - ip0]^{-1}$ is nothing but the Green's or system function of the linear harmonic undamped oscillator, $\tilde{S}_0(p, \omega^2) = \lim_{\gamma \to 0} FS_\gamma(t)$ with $\ddot{S}_\gamma(t) + \gamma\, \dot{S}_\gamma(t) + \omega^2\, S_\gamma(t) = \delta(t)$, $S_\gamma(t) = 0$ for $t < 0$ (retarded function), the relation (3.10) says that the system function $\tilde{S}(p)$ of any system can be represented in terms of a (continuously infinite) set of undamped oscillator functions $\tilde{S}_0$. In particular, a discrete, lumped RCL system has this property, i.e., it can be expressed in terms of elementary LC oscillators alone.

These results are weak in the sense that, as a consequence of exploiting only the negative (!) causality axiom, i.e., the analytic structure of $\tilde{S}(k)$, there survives the unknown term $u_0(p)$. Things become definite, however, if we stick to the unsubtracted dispersion relations discussed in §2.3, obtainable by means of our residue procedure (say, by analytic continuation of spectral functions). Then we obtain instead of (3.10) the result (cf. (2.13), (2.15))

$$(3.11) \qquad \tilde{S}(p) = \operatorname*{Res}_{z=0} \left[\frac{1}{z} \int_0^\infty \frac{\rho(\omega^2)(a^2\omega^2)^z\, d\omega^2}{\omega^2 - p^2 - ip0} \right]$$

containing only the single arbitrary scaling parameter a^2. This is justified by recalling that in the case of a causal and singular system with (3.4), $s(t) \sim t^{-\beta}$, $\beta > 0$, $t \to 0$, one may define $S = \theta(t) \circ t^{-\beta}$ by continuation of $\theta(t) \cdot t^\lambda$ in λ by starting from $\operatorname{Re} \lambda > 0$. The point here is that in this way the familiar subtraction difficulties, i.e., the indeterminacy arising when a causal system is singular ($S = 0$ for $t < 0$ and singular as $t \to +0$, respectively, $\tilde{S}(k) \to \infty$ as $|k| \to \infty$) can be overcome.

At this occasion we wish to recall the high frequency bound discussed in §2.4. One can prove this bound directly for $\tilde{S}(p)$, i.e., if the system is causal ($S(t) = 0$ for $t < 0$), then $\tilde{S}(k)$ satisfies the relation

$$(3.12) \qquad |S(k)| \leqq Ce^{|k|^\alpha}, \qquad |k| \to \infty, \quad \alpha < 1.$$

In terms of the considerations carried through in §2.4, one may say this: If one looks for $S(t)$ only in a neighborhood of $t = 0$, say, $S(t) \approx \sum_0^\infty a_n \delta^{(n)}(t)$, then the coefficients a_n must be such that this $S(t)$ maintains its support localized at $t = 0$, i.e., this support should not extend to the region $t < 0$. This means that the a_n satisfy (2.21) from which $|a_n| \leqq (n!)^{-1-\epsilon}$ and, therefore, (3.12) follows.

A similar analysis may be performed in the case of the admittance $Y(t)$. From the positive-definiteness condition (3.6) together with the causality postulate ($Y(t) = 0$ for $t < 0$), it follows that $\tilde{Y} = FY$ is positive real, i.e., $\tilde{Y}(k)$ is a Herglotz function and, therefore, satisfies the Cauer representation

$$\tilde{Y}(k) = Ck + \frac{k^2}{\pi} \text{ p.v. } \int_{-\infty}^{\infty} \frac{\operatorname{Im} \tilde{Y}(k')\, dk'}{k'^2(k' - k)} \tag{3.13}$$

($\tilde{Y}$ may increase no faster than $|k|$ as $|k| \to \infty$). The term Ck is a "subtraction" term, a particular case of (1.44). The relation (3.13) follows of course from the indicatrix formalism (§1.3) or, alternatively, by applying Cauchy's formula to $\tilde{Y}(k)/k^2$. If, however, one does not only appeal to the causality axiom (which is a negative statement telling only what does not happen (acausality)) but makes a direct attack on the singularity that $S(t)$ and $Y(t)$ possess at $t = 0$ in some systems (e.g., in the case of a transmission line), then, by our previous techniques, we get instead of (3.13) the representation

$$\tilde{Y}(k) = \frac{1}{\pi} \operatorname*{Res}_{z=0} \left[\frac{1}{z} \int_{-\infty}^{\infty} \frac{\operatorname{Im} \tilde{Y}(k')\, (ak')^z\, dk'}{k' - k} \right] \tag{3.14}$$

containing the arbitrary scaling parameter a. In physical terms, the constant C in (3.13) represents the capacity of the system and one sees from a comparision of (3.13) and (3.14) that the scaling parameter a is equivalent to C ($C \cong \log a$ or so). a fixes, as we have seen in §2.1, the time or frequency unit of the system and so does, we conjecture, the capacity C. Since, as we know from §2.1, a is related to the dilatation group, we would like to know from the network theorists what C has to do with dilatation symmetry.

The Cauer representation (3.13) and its generalization (3.14) says, as do (3.10), (3.11), that the admittance of any physical system can be represented by a continuously infinite sum of undamped harmonic oscillator admittance $\tilde{Y}_0(k, \omega^2)$. Thus a discrete RCL system is equivalent to a continuous LC system (a rather surprising fact as we believe). These questions should be further investigated.

3.2. Unstable systems and essential singularities. From (3.12), $\lim_{|k|\to\infty} |\tilde{S}(k)| \exp(-\epsilon |k|) = 0$, it follows at once that possible essential singularities of $S(t)$ are of the type $S(t) \approx \exp(1/(At^\beta))$, $\beta > 0$, as $t \to 0$. There are also systems with admittances of the type $\exp(-1/(Ak))$. Systems with $f(x) \approx \exp(1/(Ax^\lambda))$, $\lambda > 0$, appear to yield discontinuous functionals since one can easily construct functions $\varphi_m(x) \in D$ with $\varphi_m^{(n)}(0) = 0$, $n = 0, 1, \cdots$, such that $f\langle\varphi_m\rangle$ does not tend to zero if the sequence $\varphi_m(x)$ tends uniformly to zero. However, any $f = \exp(1/(Ax^\lambda))$ with $\lambda > 0$ must be viewed as a generalized function on S_0^β (with $\|\varphi\|_p = \sup_q \max_x |\varphi^{(q)}|$

$\cdot(B+\rho)^q q^{q\beta})$. Expanding $\exp(1/(Ax^\lambda))$ one can show [6] that the series converges for $\beta < 1 + 1/\lambda$ so that the functional is continuous in $S_0'^\beta$. Systems of this type, however, appear to be unstable since the behavior of f for $A > 0$ is radically different from the one with $A < 0$. Examples of this type are discussed in quantum mechanical terms in §4.3. One would like to know from the network theorists which role is played in system theory by system functions possessing essential singularities.

4. Mathematical problems of elementary particle physics. Elementary particle theories presently make use of two schemes, relativistic quantum field theory and analytic S-matrix theory (bootstraps, etc.). We shall discuss here a few mathematical aspects of these formalisms.

4.1. Quantum field theory. In this theory, interacting particles are described by quantized interacting fields $A(x)$ supposed to be operator-valued generalized functions, symbolically, $A\langle\varphi\rangle = \int_{-\infty}^{\infty} A(x)\varphi(x)\, d^4x$, acting on a Hilbert space of states according to

$$\Psi = a_0\Psi_0 + a_1\int_{-\infty}^{\infty} A(x)\varphi(x)\Psi_0\, d^4x \tag{4.1}$$
$$+ a^2 \iint_{-\infty}^{\infty} A(x)A(y)\,\varphi(x)\varphi(y)\,\Psi_0\, d^4x d^4y + \cdots,$$

where Ψ_0 is the vacuum state. The system of interacting particles is supposed to satisfy axioms of similar nature as the ones imposed on passive systems in §3. Denoting by x a point in Minkowski space R^4, $x = (x_\nu) = (x_0, \mathbf{x})$ with the metric $x^2 = x_0^2 - \mathbf{x}^2$ and introducing the Fourier operator $Ff(x) = \int_{-\infty}^{\infty} f \exp ikx\, d^4x = \tilde{f}$ with $k = (k_0, \mathbf{k})$, $k^2 = k_0^2 - \mathbf{k}^2$, $kx = k_0x_0 - \mathbf{kx}$, $d^4x = dx_0\, d\mathbf{x}$, one requires (i) that the metric in Hilbert space be positive-definite, i.e., that $(\Psi, \Psi) \geqq 0$. In the simplest case where all the a_i vanish except a_1, we obtain from the above expansion of Ψ the positive-definiteness condition (1.45),

$$(\Psi, \Psi) = i\Delta_+'\langle\varphi * \hat{\varphi}\rangle \tag{4.2}$$
$$= i\iint_{-\infty}^{\infty} \Delta'(x_+ - y)\varphi(x)\overline{\varphi(y)}\, d^4x\, d^4y \geqq 0,$$

where the vacuum expectation value

$$i\Delta_+'(x-y) = (\Psi_0, A(x)A(y)\Psi_0) \tag{4.3}$$

is called Wightman's two-point function [17], [18]. Here we have incorporated the postulate of translation invariance or, more generally, the postulate (ii) of Lorentz invariance. Bochner's theorem (§1.3) then states that $iF\Delta_+'(x) = i\tilde{\Delta}_+'(k) \geqq 0$ is a positive measure. Although (4.2) is similar in appearance to the positive-definiteness (passivity) condition of passive system theory, we do not know what this condition here corresponds to in system theory, since in field theory the unitarity condition (3.5) is implied by the interpretation of the states Ψ in Hilbert space. The postulate (iii), positive energy spectrum of the particles, means that $k^2 = k_0^2 - \mathbf{k}^2 \geqq 0$, $k_0 \geqq 0$. Hence, with some spectral function $\rho(k^2) \geqq 0$ we have that

$$(4.4) \qquad i\tilde{\Delta}_+'(k) = i\rho(k^2)\theta(k_0)\theta(k^2) \geqq 0.$$

Following Wightman, a field theory can be formulated in terms of kernels such as Δ_+' rather than in terms of the original Hilbert space concepts. We shall say, that a particle is a free elementary particle if Δ_+' is minimal, that is to say, if any other Lorentz-invariant kernel Δ_+^*, majorized by Δ_+' in the sense that $(\Delta_+' - \Delta_+^*)\langle\varphi * \hat{\varphi}\rangle \geqq 0$, is proportional to Δ_+', then $\Delta_+^* = \alpha\Delta_+'$; i.e., if $\rho^* \leqq \rho$ implies that $\rho^* = \alpha\rho$. This implies that $\rho(k^2)$ reduces to a delta function, $\rho(k^2) = \rho_0(k^2) = \delta(k^2 - m^2)$, m being the mass of the particle. With $\rho = \rho_0$, Δ_+' reduces to the free-particle function $\Delta_+'(x) = \Delta_+(x, m^2) = F^{-1}[\delta(k^2 - m^2)\theta(k_0)\theta(k^2)]$, and with $\rho(k^2) = \int_0^\infty \rho(m^2)\delta(k^2 - m^2)\, dm^2$, we obtain that $\Delta_+'(x)$ has the spectral representation

$$(4.5) \qquad \Delta_+'(x) = \int_0^\infty \rho(m^2)\Delta_+(x, m^2)\, dm^2.$$

Here, $\Delta_+(x, m^2)$ is easily seen to be the Hankel function $\Delta_+(x, m^2) = -mH_1^{(1)}(m\sqrt{z})/8\pi\sqrt{z}$ with $z = x^2 - ix_00$. In particular, if $\rho(k^2)$ increases —and this is what happens in a perturbation approach to any field theory— Δ_+' is singular, e.g., $\Delta_+'(x) = \text{const.}\ (x^2 - ix_00)^{-2-\lambda/2}$ for $\rho(m^2) = m^\lambda$, $\lambda > 0$. We wish to point out here that the above minimal principle can also be adapted to passive system theory, whereby a passive system is "elementary" if $\rho(m^2)$ in (3.9) is minimal. The elementary passive system is then simply the undamped harmonic oscillator, $\rho(\omega^2) = \delta(\omega^2 - \omega_0^2)$ in (3.10).

The postulate (iv) of causality means in field theory that the commutator $i\Delta'(x) = (\Psi_0, [A(x/2), A(-x/2)]\Psi_0)$ vanishes for space-like arguments $x^2 = x_0^2 - \mathbf{x}^2 < 0$ (no signal traveling with a velocity faster than that of light (we assume $\hbar = c = 1$)). Since, as is easily seen, $\Delta'(x) = \Delta_+'(x) - \Delta_+'(-x)$, it follows that in case of an increasing spectral function $\rho(k^2)$, the so-called retarded function $\Delta_R'(x) = -\theta(x_0)\Delta'(x)$ and the time-ordered function

$$(4.6) \qquad \Delta_F'(x) = 2i[\theta(x_0)\Delta_+'(x) + \theta(-x_0)\Delta_+'(-x)]$$

are singular at the origin of the lightcone: $\rho(m^2) = m^{2n}$ implies

$$(4.7) \qquad \begin{aligned} \Delta_F'(x) &= \text{const. } \theta(x_0)\delta^{(n+1)}(x^2), \qquad n = 0, 1, 2, \cdots, \\ \Delta_F'(x) &= \text{const. } (x^2 - i0)^{-n-2}, \qquad n = 0, 1, 2, \cdots, \end{aligned}$$

and that, even worse, one arrives at undefined products such as $\theta(x_0) \circ (x^2 - ix_0 0)^{-n-2}$. We have shown in §2 how these difficulties—which, indeed, produce the divergences of field theory—can be overcome either by sticking to the indicatrix formalism or utilizing the analytic continuation and residue techniques. We shall analyze these questions in more detail in the next paper [19] and confine ourselves here to a few remarks. The exact analogue to the spectral representations (3.10), (3.11) of system theory is now the Lehmann representation ($k = p + iq$, $q \to 0$)

$$(4.8) \qquad \tilde{\Delta}_F'(p) = 2iM(p^2) \int_0^\infty \frac{\rho(m^2)\, dm^2}{M(m^2)(p^2 - m^2 + i0)} + u_0(p^2)$$

and the unsubtracted version

$$(4.9) \qquad \tilde{\Delta}_F'(p) = 2i \operatorname*{Res}_{z=0} \left[\frac{1}{z} \int_0^\infty \frac{\rho(m^2)(a^2m^2)^z\, dm^2}{p^2 - m^2 + i0} \right].$$

The causality postulate (iv) implies that $\Delta'(x) = 0$ for $x^2 < 0$ and $|\rho(k^2)| \leqq C \exp(|k^2|^\alpha)$, $\alpha < 1/2$ as $|k| \to \infty$. This will be shown in [19]. For example, in the case of massless particles, governed by a spectral function of the type $\rho(m^2) = \sum_0^\infty a_n(m^2)^n$, in which case

$$\Delta'(x) = \sum_0^\infty a_n(n - 1)!\ \epsilon(x_0)\delta^{(n)}(x^2),$$

it follows in analogy with §2.4 that $\Delta'(x)$ has its support on the lightcone $x^2 = 0$, and the causality condition is satisfied if and only if the a_n satisfy the condition $\lim \sup_{n\to\infty} \sqrt[n]{|a_n|(2n)!} = 0$. This yields precisely the high energy bound given above.

We wish to remark that in deriving (4.8) by means of the indicatrix techniques one has to exclude the point $x_0 = \mathbf{x} = 0$, i.e., the origin of the lightcone $x^2 = 0$, for $\Delta_F'(x)$ is the boundary value of a nonanalytic function, e.g., $\Delta_F' = (x^2 - i0)^{-n} = \lim_{\eta\to 0}(x^2 - i\eta_{\mu\nu}x^\mu x^\nu)^{-n}$, $n = 2, 3, \cdots$ ($\eta_{\mu\nu}x^\mu x^\nu$ positive-definite), singular at $x_0 = \mathbf{x} = 0$. It is clear that a detailed discussion of the theory of Lorentz-invariant generalized functions is required to deal with singular quantum field theories. But we see also that by means of the concepts devised in §2 one arrives at a formulation of quantum field theory which, owing to the missing subtractions (cf., e.g., (4.9) as confronted with (4.8)), is superior to the conventional formulation. The only thing we should point out here is that the scaling parameter a appears in quantum field theory precisely in those instances where the naive formulation produces logarithmic divergences. Or else, a stands exactly at those places where a cutoff would stand although a is by no means a cutoff. For further details, see [19].

4.2. Analytic S-matrix theory. In analytic S-matrix theory one describes the interaction, say, the mutual scattering, of two particles in terms of a scattering amplitude $T(s)$ where s is the center of mass energy squared of the two particles. Similarly as in §3, the S-matrix element S is related to T by $S = 1 + 2iT$ and, in slight extension of the results of §3, the amplitude $T(s)$ is analytic in the complex s-plane cut along the positive real axis and along part of the negative axis. In case of elastic processes, $T(s)$ satisfies the unitarity condition $\operatorname{Im} T = \sigma \mid T \mid^2$, $s \geqq 0$ (cf. also §3), where σ is a given function of s which we shall set equal to a constant g without any restriction. Let us do a kind of Wiener-Hopf decomposition of $T(s)$, $T(s) = N(s)/D(s)$, where $N(s)$ is analytic in the s-plane with only the left-hand cut L of $T(s)$, whereas $D(s)$ is analytic with only the right-hand cut R of $T(s)$ (cf. Fig. 2). Using the fact that $N(s)$ and $D(s)$ are real analytic, e.g., $\overline{N(s)} = N(\bar{s})$, it follows by applying Cauchy's integral formula to $N(s)$ and $D(s)$ with the contours of integration indicated in Fig. 3, that $N(s)$ and $D(s)$ satisfy the equations ($D \to 1$ as $s \to \infty$) [20]

$$N(s) = \frac{1}{\pi} \int_{-\infty}^{s_0} \frac{\operatorname{Im} T(s') D(s')\, ds'}{s' - s - i0}, \qquad D(s) = 1 + \frac{g}{\pi} \int_0^{\infty} \frac{N(s')\, ds'}{s' - s - i0},$$

whence

$$N(s) = F(s) + g \int_0^{\infty} \frac{(F(s) - F(s'))N(s')\, ds'}{s' - s - i0}, \tag{4.10}$$

where

$$F(s) = \frac{1}{\pi} \int_{-\infty}^{s_0} \frac{\operatorname{Im} T(s')\, ds'}{s' - s - i0} \tag{4.11}$$

represents the force between the scattering particles. This is similar to

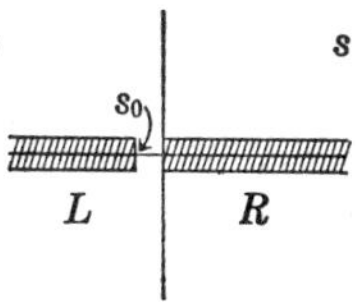

FIG. 2. *The cuts in the s-plane of the amplitude $T(s)$*

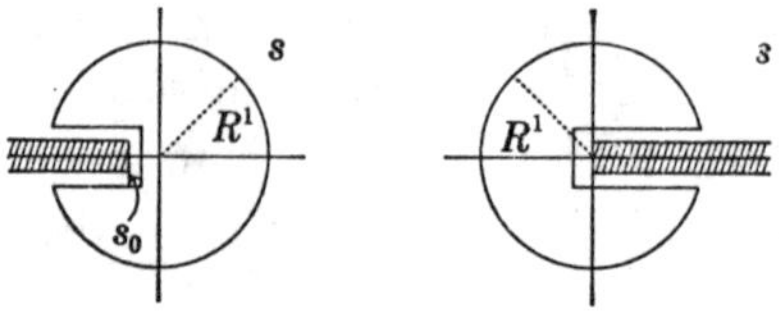

FIG. 3. *The cuts and integration contours for the functions* $N(s)$ *and* $D(s)$ $(R^1 \to \infty)$

§3 where Im $\tilde{S}(p)$ corresponding to Im $T(s)$ characterizes the dynamics of the system (and must be given at the outset).

In physics, one is, especially in case of interactions produced by the exchange of particles with higher spins and in weak interactions, confronted with the fact that Im $T(s)$ increases as $s \to \infty$ on the left-hand cut, e.g., Im $T(s) = cs^{\beta}$, $\beta \geqq 0$. Hence, (4.11) has no meaning unless one interprets the symbolic integral in terms of generalized functions (§2.3), which produces a well-defined $F(s)$ that, however, increases too as $s \to \infty$, e.g., $F(s) = \text{const. } s^{\beta}$, $\beta \geqq 0$. Hence (4.10) is a singular integral equation (not of Fredholm type) and a formal iteration yields the meaningless result $N(s) = F(s) + g\infty + g^2\infty^2 + \cdots$, where by "$\infty^n$" we mean $\lim_{K\to\infty} K^n$ obtained by cutting off the integral (4.10) at some $K > 0$. How to give a meaning to such equations will be shown in §4.3.

4.3 Singular integral equations. A situation similar to the one found in the preceding section is encountered when one looks at the Schrödinger equation for a particle with wave function $\psi(r)$ $(r \geqq 0)$ scattered by a singular potential $V(r)$ of the type

$$V(r) = gr^{-m}, \qquad m > 3. \tag{4.12}$$

The Schrödinger equation reads (g is a coupling constant)

$$\psi(r) = \varphi(r) + \int_0^{\infty} G(r, r')V(r')\psi(r')\,dr' \tag{4.13}$$

with the Green's function

$$G(r, r') = u(r)v(r')\theta(r - r') + w(r)t(r')\theta(r' - r), \tag{4.14}$$

where u, v and w, t are regular at $r = 0$, say, $u = -1/r$, $v = r'^2$, $w = -1$, $t = r'$. The equation is singular and not of Fredholm type, and a formal iteration, $\psi = \sum_0^{\infty} g^n\psi_n$, $\psi_0 = \varphi$, $\psi_{n+1} = \int_0^{\infty} GV\psi_n dr'$, $n \geqq 0$, yields again a meaningless result $\psi = \varphi + g\infty + g^2\infty^2 + \cdots$ as above.

If $g > 0$ (repulsive potential), one can show that (4.13) has an integrable function $\psi(r)$ as a solution with $d^n\psi(r)/dr^n = 0$ for $r = 0$, $n = 0, 1, 2, \cdots$.

Here it is assumed that (4.13) is interpreted in the sense of classical analysis. However, this solution in general has a singularity at $g = 0$ (besides the essential singularity at $r = 0$). This is why the formal power series expansion (iteration) does not work. For example, with $m = 4$, $\varphi = 1$, the classical solution is given by $\psi = \psi_{cl}(r) = \exp(-\sqrt{g}/r)$.

We indicate how such singular solutions can be obtained by means of generalized function techniques. Consider besides (4.13) the equation

$$\psi_1(r) = \varphi(r) + Pf \int_0^\infty G(r, r')V(r')\psi_1(r')\,dr', \tag{4.15}$$

where Pf is our "finite part" operator (§1.3) which we assume to operate on each singular term in a formal iteration of ψ_1,

$$\psi_1 = \sum_0^\infty g^n \psi_{1,n}, \qquad \psi_{1,0} = \varphi, \qquad \psi_{1,n+1} = Pf \int_0^\infty GV\psi_{1,n}\,dr', \qquad n \geqq 0.$$

Let us assume that the series $\psi_1 = \sum_0^\infty g^n \psi_{1,n}$ converges (it does so in the sense of generalized functions!). The difference $\psi_0 = \psi - \psi_1$ satisfies the equation $\psi_0 = uA + Pf \int_0^\infty GV\psi_0\,dr'$, where $A = \int_0^r v(r')V(r')\psi(r')\,dr'$ $- Pf \int_0^r v(r')V(r')\psi(r')\,dr'$ is easily shown to be independent of r since Pf commutes with differentiation. Since A is a constant, we can set $\psi_0 = A\psi_2$ obtaining for ψ_2 the equation $\psi_2 = u + Pf \int_0^\infty GV\psi\,dr'$ which (by definition of Pf) can be iterated in the same way as (4.15), $\psi_2 = \sum_0^\infty g^n \psi_{2n}$. Thus, the solution of (4.13) is given by $\psi = \psi_1 + A\psi_2$, and since $\psi^{(n)}(0) = 0$, we find that $A = -\lim_{r\to 0}[\psi_1(r)/\psi_2(r)]$. A exists if $g > 0$. Both ψ_1, ψ_2 are analytic functions of g, but A and, therefore, ψ is not. One can easily verify by means of the formula of §1 and §2 that the above procedure yields exactly the solution $\psi = \exp(-\sqrt{g}/r)$ in case of $m = 4$, $\varphi = 1$. For a more detailed discussion of this iterative treatment of singular, non-Fredholm equations, see [14].

However, one may also interpret the original equation from the outset as an integral equation in the sense of generalized functions, by regarding $V(r)$ not as a classical function but as the generalized function $Pf\,V(r)$. Then (4.13) turns into (4.15) which, in the case of $m = 4$, $\varphi = 1$, has the generalized function solution $\psi_1 = \cosh(\sqrt{g}/r) = \sum_0^\infty g^n(Pf\,r^{-2n})/(2n)!$ $\in S_\alpha'^{\beta}$, viz., a nonintegrable generalized function.

Which of the two interpretations of such singular equations is the "right" one depends on the physical situation. In the case of the quantum mechanical scattering by a repulsive singular potential ($g > 0$) the classical interpretation applies. (As a probability amplitude, $\psi(r)$ must be integrable.) In the relativistic analogue to the above equations the generalized function interpretation (i.e., (4.15)) applies (see [9]). It is not known which interpretation should be adopted if the potential $V(r)$ is attractive ($g < 0$). The reason is that such systems are unstable.

The equation (4.10) of the preceding section likewise admits a classical and a generalized function interpretation, leading to locally integrable solutions and to generalized function solutions blowing up exponentially (e.g., $N(s) \to \exp \sqrt{gs}$) as $s \to \infty$, respectively. The latter interpretation consists of considering every term in an expansion of $N(s)$ in powers of g defined by means of the generalized Hilbert transform discussed in §2.3. We do not know which of these two interpretations is the physically "right" one.

5. Perspectives on future developments. The mathematician has certainly discovered concepts and problems in the preceding sections which deserve further development, while the physicist or engineer perhaps has realized that most of his needs can be met by utilizing generalized function techniques. There are a number of questions to which we would like to call particular attention.

Our various approaches made it plausible that an effective functional calculus for generalized functions exists yielding, for example, well-defined products of distributions (free of "subtractions"), generalized Hilbert transforms for increasing functions, and giving a meaning to quite unfamiliar singular integral equations. The latter correspond to singular differential equations such as

$$\frac{d^2}{dr^2}\psi(r) + \frac{g}{r^m}\psi(r) = 0, \qquad m > 2, \tag{5.1}$$

and

$$\Box\Psi(x) + g\cdot(x^2 - i0)^{-n}\Psi(x) = 0, \qquad n \geqq 1, \tag{5.2}$$

($\Box$ = d'Alembertian) with singular generalized functions as coefficients. They represent unstable systems characterized by generalized functions with essential singularities. For example, if g in (5.1) is negative, the corresponding classical orbits of the particle are unstable. The similar situation occurs in (4.10) with the generalized function solution $N(s) = \sum_0^\infty g^n N_n(s)$, $N_0 = F$, where

$$N_{n+1}(s) = \operatorname*{Res}_{z=0}\left[\frac{1}{z}\int_0^\infty \frac{(F(s) - F(s'))N_n(s')(as')^z\, ds'}{s' - s - i0}\right]. \tag{5.3}$$

Almost nothing is known about such singular equations apart from some results discussed in the sequel to this paper [19]. Their discussion and definition requires simultaneously products of generalized functions ($\psi(r)$ is singular if $V = gr^{-m}$ is so in (5.1) !), generalized Hilbert transforms, a generalized function formulation of Wiener-Hopf techniques as the ones used in §4.2 and also should lead to some better understanding of the physical meaning of the scaling parameter a entering both the defining equations and their solutions. We consider these problems and their final solutions to be of considerable interest not only for physics but also for future developments relating to a nonlinear generalized function analysis.

Acknowledgments. The author wishes to express his sincere gratitude to the organizers of the SIAM 1966 Fall Meeting for the kind invitation and to A. S. Wightman, A. Jaffe, A. H. Zemanian, R. W. Newcomb, H. Bremermann and to many other participants for valuable discussions. Thanks are due to the Deutsche Forschungsgemeinschaft for a research grant. The kind hospitality of the Research Center and Department of Physics, New Mexico State University, Las Cruzes, New Mexico is gratefully acknowledged. Thanks are also due to Miss L. A. Needham and Mrs. B. Staffeldt for the efficient typing of the preprint of this paper.

REFERENCES

[1] L. Schwartz, *Théorie des Distributions*, vols. I and II, Hermann, Paris, 1950–51.

[2] I. M. Gel'fand, G. E. Shilov and Ya. Vilenkin, *Generalized Functions I–IV*, Academic Press, New York, 1964–1967.

[3] H. Bremermann, *Distributions, Complex Variables and Fourier Transforms*, Addison-Wesley, Reading, Massachusetts, 1965.

[4] A. H. Zemanian, *Distribution Theory and Transform Analysis*, McGraw-Hill, New York, 1965.

[5] E. J. Beltrami and M. R. Wohlers, *Distributions and Boundary Values of Analytic Functions*, Academic Press, New York, 1966.

[6] W. Güttinger, *Generalized functions and dispersion relations in physics*, Fortschr. Physik, 14 (1966), pp. 483–602.

[7] W. Güttinger and E. Pfaffelhuber, *Dynamics of unrenormalizable interactions in Minkowski and Euclidean space*, Nuovo Cimento, to appear.

[8] J. Hadamard, *Lectures on Cauchy's Problem in Linear Partial Differential Equaions*, Dover, New York, 1952.

[9] W. Güttinger and A. Rieckers, *Spectral representations of Lorentz-invariant distributions*, Comm. Math. Phys., to appear.

[10] W. Güttinger, E. Pfaffelhuber and A. Rieckers, *Lorentz-invariant generalized functions*, Nuovo Cimento, to appear.

[11] W. Güttinger, *Products of improper operators and the renormalization problem of quantum field theory*, Progr. Theoret. Phys., 13 (1955), pp. 612–626.

[12] H. König, *Multiplikation von Distributionen*, Math. Ann., 128 (1955), pp. 420–452.

[13] ———, *Multiplikationstheorie der verallgemeinerten Distributionen*, Bayer. Akad. Wiss. Math.-Natur. Kl. Abh. (N.F.), no. 82, 1957, 80 pp.

[14] W. GÜTTINGER, R. PENZL AND E. PFAFFELHUBER, *Peratization of unrenormalizable field theories*, Ann. Physics, 33 (1965), pp. 246–271.

[15] M. SCHÖNBERG AND C. BRAGA, *Formal series in quantum mechanics*, An. Acad. Brasil. Ci., 31 (1959), pp. 333–347.

[16] J. KOREVAAR, *Pansions and the theory of Fourier transforms*, Trans. Amer. Math. Soc., 91 (1959), pp. 53–101.

[17] A. S. WIGHTMAN, *Quantum field theory in terms of vacuum expectation values*, Phys. Rev., 101 (1956), pp. 860–868.

[18] S. SCHWEBER, *Introduction to Relativistic Quantum Field Theory*, Harper and Row, New York, 1962.

[19] W. GÜTTINGER AND E. PFAFFELHUBER, *Unrenormalizable field theories in terms of Lorentz-invariant generalized functions*, this Journal, 15 (1967), *pp.* 1030–1045.

[20] M. JACOB AND G. F. CHEW, *Strong Interaction Physics*, Benjamin, New York, 1964.

[21] A. P. CONTOGOURIS AND D. ATKINSON, *Solutions of singular N/D equations*, Nuovo Cimento, 39 (1965), pp. 1082–1121.

A DISTRIBUTIONAL APPROACH TO TIME-VARYING SENSITIVITY*

R. W. NEWCOMB AND B. D. O. ANDERSON†

Man, too acute, should perceive
That sensitive hearts have in grown
What's created though varied by time;
Systems are so by construct
But, as with man, little known.
Non sensed, though, in fulfilling man
A theory may guide and conduct.

1. Introduction. The theory of distributions [1], [2] has found wide application in various fields of science as, for example, in relativistic quantum mechanics [3], interaction and scattering of elementary particles [4], and network theory [5], [6], [7], [8]. Still, although results are available concerning systems analysis on a distributional basis [9], [10], little use of the rigorous theory of distributions has been made in the area of control system design. Here we investigate one of the fundamental concepts of control systems, that of sensitivity, obtaining results needed for optimal control design [11] in terms of distributions.

One of the classical problems of control theory is to reduce by feedback the sensitivity of a system to variations in the parameters of the plant. As a consequence a rather extensive literature is available concerning pertinent concepts [12], but little which directly discusses time-variable, as opposed to adjustable parameter, systems. Still time-varying, multiple-input, multiple-output systems are appearing in practical environments, by force of circumstances or as a result of implementing an optimal control law. In terms of distributional kernels we here investigate the question of when the sensitivity performance of such time-variable systems is improved by feedback.

The investigation follows the ideas of Cruz and Perkins as applied to time-invariant systems [13], and later extended by them to cover some time-variable cases [14], by considering the change in the closed loop response versus a change in the open loop response due to plant parameter changes and with the plant input held fixed. The relation between these open and closed loop response changes is linear and, for physical systems,

* Received by the editors July 7, 1966, and in revised form November 4, 1966. Contributed at the Symposium on "The Applications of Generalized Functions" sponsored by the Air Force Office of Scientific Research at the 1966 Fall Meeting of Society for Industrial and Applied Mathematics held at the State University of New York at Stony Brook, September 12–14, 1966.

† Stanford Electronics Laboratories, Stanford, California. This work was sponsored by the Air Force Office of Scientific Research, Office of Aerospace Research, United States Air Force, under Grant AF-AFOSR 337-63.

describable by a distributional kernel, the sensitivity matrix. The main result is that for sensitivity improvement through the application of feedback the sensitivity matrix must be an antecedal contraction mapping of square-integrable vectors into square-integrable vectors. Such a sensitivity matrix is analogous to the scattering matrix of a passive network, and, consequently, many of the results of passive network theory [15] apply to sensitivity problems. This analogy is used to prove the main results, and the paper can then be considered as a solidification, through a complete statement of results, and a generalization, through distribution theory, of [14].

In §2 we review the necessary distributional background with emphasis placed upon distributional kernels. In §3 we discuss the sensitivity concept introducing the sensitivity matrix as well as the return-difference. In §4 the required properties of the sensitivity matrix needed for sensitivity improvement with the application of feedback are discussed, these being obtained by the abovementioned network analogy. For convenience we adhere as closely as possible to the notation of Cruz and Perkins [13].

2. Preliminaries. Here we review and introduce those concepts associated with distributional kernels which are necessary to the sequel. Along with this we discuss the physical constraints placed on kernels used in control theory. We assume as known the basic rudiments of distribution theory [1], [2].

Let $\mathfrak{D}$, $\mathfrak{D}_+$, $\mathfrak{L}_2$, and $\mathfrak{D}'$ denote the spaces of real-valued n-vectors in one real variable with entries which are, respectively, infinitely differentiable functions zero outside a bounded set (i.e., with compact support), infinitely differentiable functions zero until a finite value of the variable (i.e., with support bounded on the left), square-integrable functions on $(-\infty, \infty)$, and distributions. The scalar product between any $y \in \mathfrak{D}'$ and $\varphi \in \mathfrak{D}$ is denoted by $\langle y, \varphi \rangle$ which, on letting $t = \infty$, is the analogue of

$$\langle y, \varphi \rangle_t = \int_{-\infty}^{t} \tilde{y}(\lambda)\varphi(\lambda)\, d\lambda, \tag{2.1a}$$

defined, for instance, when $y, \varphi \in \mathfrak{D}_+$; here the superscript tilde denotes matrix transposition. When defined we also write

$$\| y \|_t^2 = \langle y, y \rangle_t, \tag{2.1b}$$

$$\| y \| = \| y \|_\infty, \tag{2.1c}$$

and observe that $\|\cdot\|$ serves as a norm for the Hilbert space $\mathfrak{L}_2$. The norm of a bounded linear transformation $T[\ ]$ of $u \in \mathfrak{L}_2$ into $T[u] \in \mathfrak{L}_2$ is defined in the customary manner as

$$\| T \| = \sup_{\| u \| = 1} \| T[u] \|. \tag{2.2}$$

By a *distributional kernel* $k(t, \tau)$ is meant an $n \times m$ matrix of real-valued distributions in two real variables [16, p. 221]. Any linear continuous map of (m-vectors) $u \in \mathfrak{D}$ (strong topology) into (m-vectors) $y \in \mathfrak{D}'$ (weak topology) defines a distributional kernel k [17, Kernel Theorem, p. 143]:

$$y = k \bullet u. \tag{2.3a}$$

Conversely, any distributional kernel defines such a map. If we denote the scalar product in two variables by $\langle\langle , \rangle\rangle$, (2.3a) is made precise by the definition [16, p. 221] of k as that distribution which represents the map, for all $u, \varphi \in \mathfrak{D}$, through the equations

$$\langle k \bullet u, \varphi \rangle = \sum_{i=1}^{n} \sum_{j=1}^{m} \langle \langle k_{ij}(t, \tau), u_j(\tau) \rangle, \varphi_i(t) \rangle \tag{2.3b}$$

$$= \sum_{i=1}^{n} \sum_{j=1}^{m} \langle \langle k_{ij}(t, \tau), \varphi_i(t) u_j(\tau) \rangle \rangle, \tag{2.3c}$$

where, of course, k_{ij}, u_j, φ_i are the entries of k, u, φ. Applying another kernel h to y of (2.3a) we obtain

$$z = h \bullet y = h \bullet (k \bullet u) = (h \circ k) \bullet u, \tag{2.4}$$

which serves to define [16, p. 229] the Volterra *composition* $h \circ k$ of h and k as the unique kernel mapping u into z, whenever such a mapping exists. Although $h \circ k$ cannot always be formed, we note that it does exist and maps $\mathfrak{D}_+$ into $\mathfrak{D}_+$ whenever h and k both map $\mathfrak{D}_+$ into $\mathfrak{D}_+$. The composition of a number of kernels is not necessarily associative, but a sufficient condition guaranteeing associativity is that all kernels map $\mathfrak{D}_+$ into $\mathfrak{D}_+$ [18, p. 120]. With δ the unit impulse and 1_n the $n \times n$ identity matrix, $\delta 1_n = \delta(t - \tau) 1_n$ acts as the identity map under composition and hence can be composed with any kernel.

In the standard manner one defines the inverse k^{-1} under composition by

$$k^{-1} \circ k = k \circ k^{-1} = \delta 1_n . \tag{2.5}$$

Depending upon the domain of definition considered one kernel may have several inverses. Consequently, we will assume, unless otherwise stated, that if k is a mapping of $\mathfrak{D}_+$ into $\mathfrak{D}_+$ then k^{-1} is also a mapping of $\mathfrak{D}_+$ into $\mathfrak{D}_+$. For such a mapping (2.5) means that for any $u \in \mathfrak{D}_+$, $k^{-1} \bullet (k \bullet u) = (k^{-1} \circ k) \bullet u = u$.

For intuitive reasoning it is convenient to recall the functional meaning of $\bullet$ and $\circ$:

$$y = k \bullet u = \int_{-\infty}^{\infty} k(t, \lambda) u(\lambda) \, d\lambda, \tag{2.6a}$$

(2.6b) $$h \circ k = \int_{-\infty}^{\infty} h(t, \lambda)k(\lambda, \tau)\, d\lambda.$$

Also in the standard manner one defines the *adjoint* k^a through

(2.7a) $$\langle u, k^a \bullet \varphi \rangle = \langle k \bullet u, \varphi \rangle$$

for all $u, \varphi \in \mathfrak{D}$. Because of (2.3) [15, §4], one readily finds

(2.7b) $$k^a(t, \tau) = \tilde{k}(\tau, t),$$

and, thus, k^a generally will not map $\mathfrak{D}_+$ into $\mathfrak{D}_+$ when k does.

Of special interest are the nonnegative kernels [3, p. 45]. By definition, a real self-adjoint distributional kernel is *nonnegative*, written $k \geqq 0$, if, for all $\varphi \in \mathfrak{D}$,

(2.8) $$\langle k \bullet \varphi, \varphi \rangle \geqq 0.$$

Turning to more physical notions, a system can be conceived as a transformation, here assumed linear, mapping inputs u into outputs y. Because we wish to treat physical systems, we can assume that $u, y \in \mathfrak{D}_+$, [19]. Further, discontinuous transformations seem physically out of the question. Consequently, since $u \in \mathfrak{D} \subset \mathfrak{D}_+$ and $y \in \mathfrak{D}_+ \subset \mathfrak{D}'$, we find by the kernel theorem that a linear physical system is described by a distributional kernel k through $y = k \bullet u$, at least for all $u \in \mathfrak{D}$. But, in actual fact, $k \bullet u$ can be extended to hold for all $u \in \mathfrak{D}_+$ since the original system transformation allows all $u \in \mathfrak{D}_+$, [16, p. 224]. Thus, as we expect also from physical arguments, $y = k \bullet u$ is defined for all $u \in \mathfrak{D}_+$, with $y \in \mathfrak{D}_+$. For some systems $y = k \bullet u$ can be defined for distributional inputs other than $u \in \mathfrak{D}_+$. For example, an extension to $\mathfrak{L}_2$ will be important in the next section while the consideration of impulsive u allows for the physical interpretation of k as an impulse response matrix.

3. The sensitivity matrix. In this section we define and interpret the sensitivity matrix. By the reasoning of §2, all impulse response matrices are assumed to be distributional kernels mapping $\mathfrak{D}_+$ into $\mathfrak{D}_+$.

Consider a fixed linear plant P which takes (m-vector) inputs u_o into (n-vector) outputs y_o and which is subject to variations in a parameter x. Then P is described by its $n \times m$ impulse response matrix p_x, a distributional kernel dependent on x. To obtain desirable transfer characteristics a controller G_1 is customarily inserted before the plant, as shown in Fig. 1, such that actual (p-vector) inputs r are modified by the $m \times p$ controller impulse response matrix g_1 to obtain the plant inputs:

(3.1a) $$y_o = p_x \bullet u_o, \qquad u_o = g_1 \bullet r,$$

or

(3.1b) $$y_o = (p_x \circ g_1) \bullet r.$$

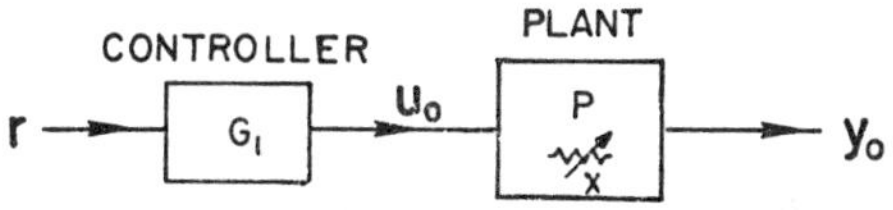

FIG. 1. *The open loop system*

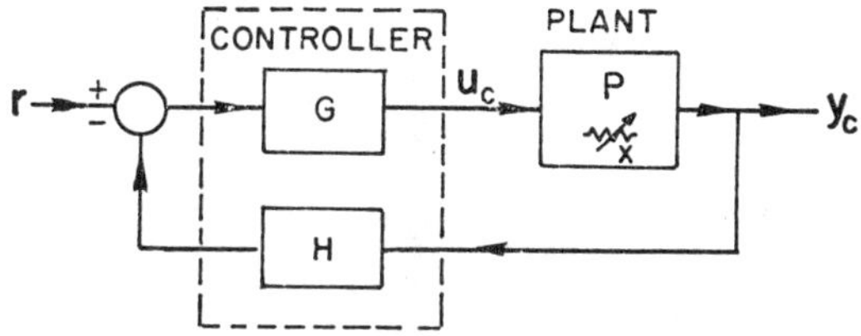

FIG. 2. *The closed loop system*

The $n \times p$ impulse response matrix of the open loop system, Fig. 1, is then $p_x \circ g_1$, and one notes that although y_o depends upon x, u_o does not since r and g_1 are assumed free of such variations. However, classical control theory [20, p. 211] recognizes that a redesign of the controller to incorporate feedback, which will cause the plant input to vary properly with x, can lead to smaller variations in the plant output with x. A general closed loop configuration of this type is shown in Fig. 2, where the controller components G and H are described by their $m \times p$ and $p \times n$ impulse response matrices g and h, also assumed independent of x. We note that

(3.2a) $$y_c = p_x \bullet u_c, \qquad u_c = g \bullet r - (g \circ h) \bullet y_c,$$

and hence, for the closed loop system,

(3.2b) $$y_c = [(\delta 1_n + p_x \circ g \circ h)^{-1} \circ p_x \circ g] \bullet r.$$

For a meaningful design the open and closed loop controllers are, of course, constructed such that the respective plant outputs are equal, $y_c = y_o$, for a given input r when the parameter x assumes its design value $x = x_d$. This entails, for $x = x_d$, that $u_c = u_o$ or, from (3.1a) and (3.2a),

(3.3) $$[g - g_1 - g \circ h \circ p_x \circ g_1] \bullet r = 0,$$

which can be used to design g and h. However, the problem of interest here is the determination of the constraints on g and h such that variations in the closed loop output y_c, due to changes in x, are smaller than the corresponding variations in the open loop output y_o, for a given g_1 and p_x.

For such an investigation, in contradistinction to Cruz and Perkins [13, p. 217], let primed quantities denote the designed situation $x = x_d$, and unprimed quantities the situation for general x; thus $p_x' = p_{x_d}$. We then introduce the *open* and *closed loop output errors*, e_o and e_c, through

(3.4a) $$e_o = y_o' - y_o,$$

(3.4b) $$e_c = y_c' - y_c.$$

Then $e_o = e_c + (y_c - y_o)$ and, from (3.1a) and (3.2b), $y_c - y_o = p_x \bullet [g - g_1] \bullet r - p_x \bullet g \bullet h \bullet y_c$, which on subtraction and addition of $p_x \bullet g \bullet h \bullet y_c' = (p_x \circ g \circ h) \bullet p_x' \bullet u_c' = (p_x \circ g \circ h \circ p_x') \bullet u_o = p_x \bullet [g \circ h \circ p_x' \circ g_1] \bullet r$, and the use of (3.3) (primed), yields

(3.5a) $$e_o = [\delta 1_n + p_x \circ g \circ h] \bullet e_c.$$

We note that the feedback factor

(3.5b) $$f = \delta 1_n + p_x \circ g \circ h$$

is the *return-difference* [21, p. 48], that is, the difference between "unit" signal applied to the controller at the input to H and the signal returned to the controller via the feedback path of Fig. 2 when $r = 0$.

Since it is of most interest to evaluate the closed loop changes in terms of the open loop ones, we define the *sensitivity matrix* s as

(3.6a) $$s = [\delta 1_n + p_x \circ g \circ h]^{-1},$$

for which

(3.6b) $$e_c = s \bullet e_o.$$

In summary, given a physical system designed with open and closed loop controllers to obtain a given output-input relationship, a linear transformation exists relating the changes in the open loop output to changes in the closed loop output, due to variations in a plant parameter x, the relationship being represented by an $n \times n$ distributional kernel s, the sensitivity matrix. Being the inverse of the return-difference matrix f, s agrees with the more classical concepts for time-invariant single-input single-output systems [22, p. 121].

4. Sensitivity improvement criteria. Here we show that the closed loop system yields improved sensitivity performance if and only if the sensitivity matrix defines an antecedal map of $\mathcal{L}_2$ into $\mathcal{L}_2$ of norm bounded by unity.

We begin by restricting to inputs r in $\mathfrak{D}_+$, in which case we know on physical grounds that $e_o, e_c \in \mathfrak{D}_+$. Consequently, through (2.1), the quadratic performance indices $\langle e_o, e_o \rangle_t$ and $\langle e_c, e_c \rangle_t$ are well defined. A reasonable criteria for improvement of sensitivity performance is then, that for any given $r \in \mathfrak{D}_+$,

(4.1) $$\mathcal{E}(t) = \| e_o \|_t^2 - \| e_c \|_t^2$$

satisfies, for all finite t,

$$\mathcal{E}(t) \geqq 0. \tag{4.2}$$

That is, we will say that *sensitivity is improved* by feedback if at each instant of time the integral of the sum of squared error components is not increased by the application of feedback.

At this point we note that, for a system in which sensitivity is improved, the situation is analogous to that for passive (linear and solvable) n-port networks. Thus, if we consider e_o as incident voltages, v^i, and e_c as reflected voltages, v^r, then s is completely analogous to the scattering matrix of the network with $\mathcal{E}(t)$ the total input energy [15]. Consequently, by choosing $e_o(\lambda) = 0$ for $\lambda < t$, we see that $e_c(\lambda) = 0$ for $\lambda < t$, from (4.1), which implies [15, §4] that s is antecedal, that is, satisfies $s(t, \tau) = 0_n$ for $t < \tau$, where 0_n is the $n \times n$ zero matrix. Further, s can be extended to map $\mathcal{L}_2$ into $\mathcal{L}_2$. To obtain this extension we note that $\| e_o \|$ is well defined for $e_o \in \mathcal{L}_2$, thus implying that $e_c \in \mathcal{L}_2$ by (4.2). A system for which sensitivity is improved therefore defines a map of $e_o \in \mathcal{L}_2$ into $e_c \in \mathcal{L}_2$, this map being represented by s for $e_o \in \mathfrak{D}_+ \cap \mathcal{L}_2$. But $e_c = s \bullet e_o$ is valid [16, p. 224] for all $e_o \in \mathcal{L}_2$, in which case $\mathcal{E}(\infty) \geqq 0$ implies that $\| e_o \| \geqq \| s \bullet e_o \|$, or what is the same, the $\mathcal{L}_2$ into $\mathcal{L}_2$ transformation defined by s has its norm bounded by unity; for notational convenience we write this result as $\| s \| \leqq 1$ since no confusion can arise even though distributions are under consideration. Omitting further particulars which are detailed elsewhere [15, §4], we then have the main result.

THEOREM. *Sensitivity is improved by feedback if and only if the sensitivity matrix s satisfies the following conditions*:

(a) s *maps* $\mathcal{L}_2$ *into* $\mathcal{L}_2$;

(b) $s(t, \tau) = 0_n$ *for* $t < \tau$;

(c) $\| s \| \leqq 1$.

One of the most useful properties that can be determined from the theorem is that $s(t, \tau)$ is a measure (i.e., at most impulsive) in both variables simultaneously over any compact set of the (t, τ)-plane [15, §4]. Another property is seen by writing (4.2) in more detail:

$$\mathcal{E}(t) = \langle e_o, e_o \rangle_t - \langle s \bullet e_o, s \bullet e_o \rangle_t \tag{4.3a}$$

$$= \langle (\delta 1_n - s^a \circ s) \bullet e_o, e_o \rangle_t. \tag{4.3b}$$

We comment that $s^a \circ s$ is a well-defined kernel mapping $\mathcal{L}_2$ into $\mathcal{L}_2$; since s maps $\mathcal{L}_2$ into $\mathcal{L}_2$ so does s^a and consequently also $s^a \circ s$. Thus, letting $t \to \infty$ with $e_o \in \mathfrak{D}$, we see that

$$R = \delta 1_n - s^a \circ s \geqq 0 \tag{4.3c}$$

or R is a nonnegative kernel. Note that, in some sense, the "smaller" R, the less the sensitivity improvement, the limit being for $s^a = s^{-1}$. In terms of the return-difference we also have, from (3.5) and (3.6),

$$(s^a)^{-1} \circ R \circ s^{-1} = f^a \circ f - \delta 1_n \geqq 0. \tag{4.3d}$$

If the system is time-invariant [10], then $s(t, \tau) = s(t - \tau, 0)$, in which case one can take the Laplace transform $\mathfrak{L}[\cdot]$ (see [23], [24]) to obtain

$$S(p) = \mathfrak{L}[s(t, 0)]. \tag{4.4}$$

Again by analogy with the network situation [15], [25, p. 116], $S(p)$ must be *bounded-real*, that is, satisfy the following corollary, where a superscript asterisk denotes complex conjugation.

COROLLARY. *If* $s(t, \tau) = s(t - \tau, 0)$, *then sensitivity is improved by feedback if and only if*:

(a) $S(p)$ *is holomorphic in* $\operatorname{Re} p > 0$;
(b) $S^*(p) = S(p^*)$ *in* $\operatorname{Re} p > 0$;
(c) $1_n - \tilde{S}(p^*)\, S(p)$ *is positive semidefinite in* $\operatorname{Re} p > 0$.

When $S(p)$ is rational this precisely states the results discussed by Cruz and Perkins [13, p. 219].

5. Discussion. By observing the strict equivalence between the scattering matrix of a passive n-port and the sensitivity matrix of an n-output system for which sensitivity is improved by feedback application, the conditions of the theorem have been obtained. The results rest heavily upon the theory of distributions for their formulation with the theorem showing, however, that no "worse" than impulses appear in s. For example, in the case of a system described by differential equations (a differential system) s takes the form

$$s(t, \tau) = A(t)\delta(t - \tau) + \Phi(t)\tilde{\Psi}(\tau)1(t - \tau), \tag{5.1}$$

where $1(\cdot)$ is the unit step function, A has eigenvalues no greater than one, and Φ and Ψ are infinitely differentiable matrices subject to the nonnegative kernel constraint of (4.3c). Although the theory does allow the consideration of any distributional kernel s mapping $\mathfrak{D}_+$ into $\mathfrak{D}_+$, the results show that for sensitivity improvement only $\mathfrak{L}_2$ maps need be considered, thus justifying the previous metric space assumption [14]. The kernel theorem and related distributional developments have allowed the complete sensitivity improvement conditions of the main result theorem.

If one has a finite dynamical (differential) system with H following the plant in the forward loop and unity feedback (i.e., Fig. 2 with y_c the output of H in place of P), then, under broad conditions, it can be shown that an "optimally" designed linear feedback law leads to sensitivity improvement [11]. Conversely, strict sensitivity improvement means that, for a time-invariant finite dynamical system, there is some quadratic loss function for which the feedback system is optimal [26]. Consequently, the results should be of some practical importance. It should, however, be pointed out that the theory of this paper is based upon creation of the system in the zero state at $t = -\infty$; nevertheless, a finite-dimensional state space is not assumed in the general arguments.

It is clear that the theory is valid for the most general linear systems of interest, but does not cover nonlinear systems, even though many of the concepts should carry over to the latter case. It is not so clear, however, that the variation of the disturbing parameter x should be "nonexistent". That is, x is essentially fixed for all time in the analysis and two "different" systems compared, one with x arbitrary and one with x at its design value x_d. This implied assumption is inherent in all such work and is physically reasonable for slow variations in x.

It is important to recognize the various extensions of kernel domains used in the theory. Although scant mention has been made of the topologies underlying the range and domain spaces, it being felt that these are of minor concern in physical situations, full justifications for the existence and extensions of the kernels can be found in Schwartz [16, p. 224]. However, the study does point out that for greater insight into sensitivity matrices a more detailed study of nonnegative distributional kernels is in order, there being very little presently available [3], [15].

Acknowledgments. The authors are grateful to J. B. Cruz, Jr., for discussion suggesting the material, W. R. Perkins for discussion on the material, and P. Kokotović for his interest and encouragement in the ideas. The care in the excellent preparation of the final manuscript by Mary Ellen Terry is likewise acknowledged, as is the reviewer's assistance in improving the readability of the paper.

REFERENCES

[1] L. Schwartz, *Théorie des Distributions*, vol. I, Hermann, Paris, 1957.

[2] ———, *Théorie des Distributions*, vol. II, Hermann, Paris, 1959.

[3] ———, *Application of distributions to the study of elementary particles in relativistic quantum mechanics*, Tech. Rep. 7, NR041-221, Department of Mathematics, University of California, Berkeley, 1961.

[4] W. Güttinger, R. Penzl and E. Pfaffelhuber, *Peratization of unrenormalizable field theories*, Ann. Physics, 33 (1965), pp. 246–271.

[5] R. W. Newcomb, *Hilbert transforms and positive-real functions*, Proc. IRE, 50 (1962), pp. 2516–2517.

[6] A. H. Zemanian, *An N-port realizability theory based on the theory of distributions*, IEEE Trans. Circuit Theory, CT-10 (1963), pp. 265–274.

[7] R. W. NEWCOMB, *The foundations of network theory*, The Institute of Engineers Australia, Electrical and Mechanical Engineering Transactions, EM6 (1964), pp. 7–12.

[8] M. R. WOHLERS AND E. J. BELTRAMI, *Distribution theory as the basis of generalized passive-network analysis*, IEEE Trans. Circuit Theory, CT-12 (1965), pp. 164–170.

[9] V. DOLEŽAL, *Dynamics of Linear Systems*, Publishing House of the Czechoslovak Academy of Sciences, Prague, 1964.

[10] R. W. NEWCOMB, *Distributional impulse response theorems*, Proc. IEEE, 51 (1963), pp. 1157–1158.

[11] B. D. O. ANDERSON, *Sensitivity improvement using optimal design*, Proc. IEE, 113 (1966), pp. 1084–1086.

[12] P. V. KOKOTOVIĆ AND R. S. RUTMAN, *Sensitivity of automatic control systems* (Survey), Avtomat. i Telemeh., 26 (1965), pp. 730–750.

[13] J. B. CRUZ, JR. AND W. R. PERKINS, *A new approach to the sensitivity problem in multivariable feedback system design*, IEEE Trans. Automatic Control, AC-9 (1964), pp. 216–223.

[14] W. R. PERKINS AND J. B. CRUZ, JR., *Sensitivity operations for linear time-varying systems*, Sensitivity Methods in Control Theory, L. Radanovic, ed., Pergamon Press, New York, 1966, pp. 67–77.

[15] B. D. O. ANDERSON AND R. W. NEWCOMB, *Functional analysis of linear passive networks*, Internat. J. Engr. Sci., to appear.

[16] L. SCHWARTZ, *Théorie des noyaux*, Proceedings of the International Congress of Mathematicians, Cambridge, Massachusetts, 1950, vol. I, American Mathematical Society, Providence, 1952, pp. 220–230.

[17] ———, *Espaces des fonctions différentiables à valeurs vectorielles*, J. Analyse Math., 4 (1954–1955), pp. 88–148.

[18] ———, *Théorie des distributions à valeurs vectorielles*, Ann. Inst. Fourier (Grenoble), 7 (1957), Chap. 1, pp. 1–141.

[19] R. W. NEWCOMB, *On the definition of a network*, Proc. IEEE, 53 (1965), pp. 547–548.

[20] H. CHESTNUT, *Systems Engineering Tools*, John Wiley, New York, 1965.

[21] H. W. BODE, *Network Analysis and Feedback Amplifier Design*, Van Nostrand, New York, 1945.

[22] J. G. TRUXAL, *Automatic Feedback Control Systems Synthesis*, McGraw-Hill, New York, 1955.

[23] L. SCHWARTZ, *Transformation de Laplace des distributions*, Comm. Sém. Math. Univ. Lund [Medd. Lunds Univ. Mat. Sem.], Tome Supplémentaire, 1952, pp. 196–206.

[24] R. W. NEWCOMB AND R. G. OLIVEIRA, *Laplace transforms-Distributional theory*, Tech. Rep. 2250–3, Stanford Electronics Laboratories, Stanford, California, 1963.

[25] D. C. YOULA, L. J. CASTRIOTA AND H. J. CARLIN, *Bounded real scattering matrices and the foundations of linear passive network theory*, IRE Trans. Circuit Theory, CT-4 (1959), pp. 102–124.

[26] B. D. O. ANDERSON, *The inverse problem of optimal control*, Tech. Rep. 6560-3, Stanford Electronics Laboratories, Stanford, California, 1966.

DISSIPATIVE OPERATORS, POSITIVE REAL RESOLVENTS AND THE THEORY OF DISTRIBUTIONS*

E. J. BELTRAMI†

1. Introduction. Among the class of linear operators on a Hilbert space H_0 which generate a strongly continuous semigroup $S(t)$, $t \geqq 0$, the maximal dissipative operators with dense domain distinguish themselves by generating *contraction* semigroups. Now the resolvents R_λ of such operators A are positive real in the sense that $(R_\lambda u, u)$ are positive real in complex λ for each $u \in H_0$ (see, for example, Dolph [1]). In Theorem 4.2 we show that knowledge that the resolvent is positive real is equivalent to saying that A is maximal dissipative with dense domain. This result is of interest in the study of passive Hilbert systems (cf. [1]) in view of the known fact that the immittance of a linear time invariant system is positive real if and only if the system is passive in the usual sense. Our proof uses the Bochner-Schwartz theorem of the theory of distributions.

In Theorem 4.1 we give another proof of the fact, already known to Dolph [1], that a linear closed operator with dense domain is maximal dissipative if and only if its resolvent is positive real and has the spectral representation $\int \frac{d\psi_\zeta}{i\zeta - \lambda}$, where ψ_ζ is a generalized resolution of the identity. To do this we first examine the distributional boundary behavior of positive real functions in §3. Finally, in §5, the close connection is observed between the immittance of passive systems and positive real resolvents.

2. Preliminaries. For later reference we itemize below certain notions from operator theory. A full discussion may be found in Hille and Phillips [2] and Phillips [3].

Let A be a linear operator (generally unbounded and nonself-adjoint) on a Hilbert space H_0 having inner product $(,)$ and denote by $\mathfrak{D}(A)$ the domain of A.

DEFINITION 2.1. A is *dissipative* if $2\text{Re}\,(Au, u) = (Au, u) + (u, Au) \leqq 0$ for all $u \in \mathfrak{D}(A)$ and A is *maximal dissipative* if it is not the proper restriction of any other dissipative operator.

DEFINITION 2.2. A one-parameter family of bounded operators $S(t)$ on H_0, $t \geqq 0$, is a *strongly continuous semigroup* if $S(t + \tau) = S(t)S(\tau)$ and $\| S(t)u - S(\tau)u \| \to 0$ as $t \to \tau$ for $t, \tau \geqq 0$, with $S(0) = I$.

---

* Received by the editors October 7, 1966. Contributed at the Symposium on "The Applications of Generalized Functions" sponsored by the Air Force Office of Scientific Research at the 1966 Fall Meeting of Society for Industrial and Applied Mathematics held at the State University of New York at Stony Brook, September 12–14, 1966.

† State University of New York at Stony Brook, Stony Brook, Long Island, New York 11790. This research was supported by the Air Force Office of Scientific Research under Grant AF-AFOSR-1154-66.

DEFINITION 2.3. A strongly continuous semigroup is a *contraction semigroup* whenever $\| S(t) \| \leqq 1$.

DEFINITION 2.4. A one-parameter family of bounded self-adjoint operators ψ_w on H_0 will be called a *generalized resolution of identity* if

(i) $\psi_w \geqq \psi_{w'}$ for $w > w'$, i.e., if $(\psi_w u, u) \geqq (\psi_{w'} u, u)$ for all $u \in H_0$,

(ii) $\psi_{w+0} = \psi_w$,

(iii) $\psi_\infty = I$, $\psi_{-\infty} = 0$.

DEFINITION 2.5. A function $f(\lambda)$ of complex λ is *positive real* if $f(\lambda)$ is holomorphic in the half-plane $\operatorname{Re} \lambda > 0$ and if it satisfies there the conditions that $\operatorname{Re} f(\lambda) \geqq 0$ and $\overline{f(\lambda)} = f(\bar{\lambda})$.

DEFINITION 2.6. For any closed linear operator A the resolvent $R_\lambda = (I\lambda - A)^{-1}$ is used to be positive real if $(R_\lambda u, u)$ is positive real in λ for each u.

We remark that the requirement that $(R_\lambda u, u)$ be real for real λ is not important for the arguments in this paper although it does play a role in passive system theory. It is inserted as a condition here only in order to complete the usual definition of positive reality.

We will need the following two results.

THEOREM 2.1. *In order that an operator A be the infinitesimal generator of a strongly continuous contraction semigroup it is necessary and sufficient that A be a closed linear operator with dense domain in H_0 whose resolvent R_λ exists for* $\operatorname{Re} \lambda > 0$ *and satisfies* $\sigma \| R_\sigma \| \leqq 1$, $\lambda = \sigma + iw$.

THEOREM 2.2. *An operator A is maximal dissipative with dense domain if and only if it generates a strongly continuous contraction semigroup.*

One final remark is that, if R_λ is the resolvent of a maximal dissipative operator A with dense domain, then, in the strong sense, $R_\lambda u = \int_0^\infty e^{-\lambda t} S(t) u \, dt$, where $S(t)$ is the semigroup associated with A. Thus,

$$(R_\lambda u, u) = \int_0^\infty e^{-\lambda t} (S(t)u, u) \, dt. \tag{2.1}$$

3. Distributional boundary values of positive real functions. Let $f(\lambda)$ be positive real, $\lambda = \sigma + iw$. Then $f(\lambda)$ admits the representation

$$f(\lambda) = A\lambda + \int \frac{i\zeta\lambda - 1}{i\zeta - \lambda} \, d\mu(\zeta), \tag{3.1}$$

where A is a nonnegative constant and μ is a bounded nondecreasing function. The formula (3.1) is due to Cauer and is a half-plane version of a theorem of Herglotz (see Cauer [4]). In particular, when $f(\lambda)$ satisfies the growth condition $\sigma f(\sigma) = O(1)$ at infinity, then one obtains the representation

$$f(\lambda) = \int \frac{d\gamma(\zeta)}{i\zeta - \lambda}, \tag{3.2}$$

where γ is a bounded nondecreasing and nonnegative measure defined by

$$\gamma(w) = \int_{-\infty}^{w} (1 + \zeta^2)\, d\mu(\zeta). \tag{3.3}$$

(For a proof of (3.2), see Shohat and Tamarkin [5, pp. 23–26].) Note that unless $f(\lambda)$ satisfies a suitable growth condition, as the one indicated above, the formula (3.2) may fail to have a sense since the measure γ is in general unbounded.

The next theorem tells us that positive real functions take on boundary values in some suitable way.

THEOREM 3.1. *Let $f(\lambda)$ be positive real. Then, as $\sigma \to 0$, $f(\lambda)$ takes on a boundary value f_w in the Schwartz topology of tempered distributions ($\mathcal{S}'$ topology); f_w is the distributional Fourier transform of some $S_t \in \mathcal{S}'$ with support in the half-axis, $t \geqq 0$, and the Laplace transform of S_t is $f(\lambda)$. Moreover, as $\sigma \to 0$, $2\mathrm{Re}\, f(\lambda) \to f_w + \overline{f_n}$ which defines the tempered and nonnegative measure given by $2\pi\gamma$ (i.e., $\langle 2\mathrm{Re}\, f(\lambda), \varphi(w)\rangle \to 2\pi \int \varphi(w)\, d\gamma(w)$ for all $\varphi \in \mathcal{S}$).*

Proof. The theorem is but a special case of a more general result. For a full discussion and proof, see Beltrami and Wohlers [6, Theorem 2.7, p. 52; Theorem 3.17, p. 86].

Remarks. Since $f_w + \overline{f_w}$ defines a nonnegative and tempered measure, the Bochner-Schwartz theorem [7, vol. 2, pp. 130–132] tells us that its inverse Fourier transform $W_t = S_t + S_{-t}$ is a nonnegative definite distribution. In particular, if W_t is a continuous function, then $W_t \leqq W_0$. Also, if γ is a bounded measure, then it can always be normalized so that $\gamma(w) \to 0$ as $w \to -\infty$ without changing the representation (3.2). In this case, $W_t = \int e^{itw}\, d\gamma(w)$ and $W_0 = \gamma(+\infty)$.

Let $S(t)$ be a strongly continuous semigroup, $t \geqq 0$. Then $(S(t)u, u)$ is continuous in $t > 0$ for $u \in H_0$. If we define S_t by $(S(t)u, u)$ for $t \geqq 0$ and zero elsewhere, then $S_t + S_{-t}$ is continuous for $t \neq 0$ with a simple jump at the origin (recall that $S_t \to (u, u)$ as $t \to 0+$). If we now set $W_t = S_t + S_{-t}$ for $t \neq 0$ with $W_0 = S_0/2$, then W_t is a continuous function which defines the same distribution as $S_t + S_{-t}$.

Our purpose in the next section is to characterize maximal dissipative operators by means of their resolvents.

4. Spectral representation of positive real resolvents. We begin by offering a different proof of a result which, in somewhat altered guise, summarizes what is crucial in Theorems 4.14, 4.15, 4.16 of [1]. Our argument is similar to that used by Bochner to establish Stone's theorem on unitary operators and is of some interest in itself.

THEOREM 4.1. *Let A be a linear closed operator with dense domain in H_0. Then a necessary and sufficient condition in order that A be maximal dissipative is that its resolvent R_λ be positive real and have the spectral representation*

$$R_\lambda(A) = \int \frac{d\psi_\zeta}{i\zeta - \lambda}, \tag{4.1}$$

where ψ_ζ is a generalized resolution of identity.

Proof. If (4.1) holds then, as with Dolph, we use the fact established by Schreiber [9] that

$$\left| \int \varphi(\zeta) d(\psi_\zeta u, v) \right| \leqq \text{l.u.b.} \mid \varphi(\zeta) \mid \| u \| \| v \|$$

for any bounded function φ and for any $u, v \in H_0$. But then, letting $\varphi(\zeta) = 1/(i\zeta - \lambda)$, we obtain l.u.b. $|\varphi(\zeta)| \leqq 1/\text{Re}\,\lambda$, and if $v = R_\lambda u$, one has

$$\| R_\lambda u \|^2 (R_\lambda u, R_\lambda u) = \int \frac{d(\psi_\zeta u, R_\lambda u)}{i\zeta - \lambda} \leqq \frac{\| u \| \; \| R_\lambda u \|}{\text{Re}\,\lambda}$$

and so, for $\sigma = \text{Re}\,\lambda > 0$,

$$\sigma \| R_\sigma \| \leqq 1. \tag{4.2}$$

Thus the resolvent of A satisfies Theorem 2.1 and hence is the infinitesimal generator of a contraction semigroup. By Theorem 2.2, A is maximal dissipative.

Conversely, by letting $v = R_\lambda u$ we obtain $u = (I\lambda - A)v$ and so, for $u \in H_0$ (the domain of R_λ is H_0), $2\text{Re}\,(R_\lambda u, u) = (R_\lambda u, u) + (u, R_\lambda u) = (v, (I\lambda - A)v) + ((I\lambda - A)v, v) = 2\text{Re}\,\{\lambda(v, v) - (Av, v)\} \geqq 2 \cdot \text{Re}\,\lambda(v, v) \geqq 0$ for $\text{Re}\,\lambda > 0$ since $\text{Re}\,(Av, v) \leqq 0$. Thus, R_λ is positive real. Also, by virtue of Theorems 2.1 and 2.2, (4.2) holds. Hence, as we saw in §3, the representation

$$(R_\lambda u, u) = \int \frac{d\gamma(\zeta \mid u)}{i\zeta - \lambda} \tag{4.3}$$

is valid where γ is a bounded nondecreasing function for each u, normalized so that $\gamma(-\infty \mid u) = 0$ and so that it is continuous from the right. We now polarize (4.3) by means of the identity $4(R_\lambda u, v) = (R_\lambda\, u + v, u + v) - (R_\lambda\, u - v, u - v) + i(R_\lambda(u + iv), u + iv) - i(R_\lambda(u - iv), u - iv)$ to obtain the representation

$$(4.4) \qquad (R_\lambda u, v) = \int \frac{d\psi(\zeta \mid u, v)}{i\zeta - \lambda},$$

where now ψ is a function of bounded variation for each u, v. Equation (4.4) is bilinear in u, v and $\psi(\zeta \mid u, u) \equiv \gamma(\zeta \mid u) \geqq 0$. Hence, by Schwarz's inequality, $|\psi(\zeta \mid u, v)|^2 \leqq \gamma(\zeta \mid u)\gamma(\zeta \mid v)$. We saw in §3 that, as $\sigma \to 0$, 2Re $(R_\lambda u, u)$ tends to $2\pi\gamma(w \mid u)$. Moreover, by the classical Bochner theorem on nonnegative definite functions,

$$\int e^{itw} d\gamma(w \mid u) = (S(t)u, u) + (S(-t)u, u),$$

which, as we saw above, is a continuous nonnegative definite function normalized so that at the origin $\gamma(\infty \mid u) = (u, u) \geqq \gamma(w \mid u)$. Thus, $\psi(\zeta \mid u, v)$ is a continuous bilinear form in u, v for each ζ. By a theorem of Riesz (see, for example, Riesz and Nagy [8, p. 202]), there exists a bounded operator ψ_ζ for each ζ such that $\psi(\zeta \mid u, v) = (\psi_\zeta u, v)$. Thus,

$$(4.5) \qquad (R_\lambda u, u) = \int \frac{d(\psi_\zeta u, u)}{i\zeta - \lambda}.$$

Since $(\psi_\zeta u, u) \geqq 0$, then ψ_ζ is self-adjoint for each ζ (see, for example, Riesz and Nagy [8, p. 229]). Also $\int d(\psi_\zeta u, u) = (\psi_\infty u, u) = (u, u)$ as we saw. Hence, $\psi_\infty = I$ and $\psi_{-\infty} = 0$. Finally, since γ is nondecreasing, $\psi_{\zeta_1} \leqq \psi_{\zeta_2}$ for $\zeta_1 < \zeta_2$ and $\psi_{\zeta+0} = \psi_\zeta$. This shows that the family ψ_ζ is a generalized resolution of the identity. Thus (4.1) is established.

We now want to show that the only way a strongly continuous semigroup, generated by an operator A, can be contractive is for the resolvent to be positive real.

THEOREM 4.2. *Let A be the infinitesimal generator of a strongly continuous semigroup $S(t)$, $t \geqq 0$. Then A is maximal dissipative with dense domain (i.e., $S(t)$ is contractive) if and only if its resolvent is positive real.*

Proof. The necessity is part of Theorem 4.1. To establish the converse, note that from Theorem 3.1, 2Re $(R_\lambda u, u)$ tends to a nonnegative and tempered measure which is the distributional Fourier transform of the nonnegative definite distribution $W(t) = (S(t)u, u) + (S(-t)u, u)$. As we saw, $W(t)$ is continuous and bounded by its value at the origin. Thus, $W(t) \leqq (u, u)$. It follows that for $t \geqq 0$, $\| S(t) \| = \text{l.u.b.}_{u \neq 0} (S(t)u, u) / (u, u) \leqq 1$ and so $S(t)$ is a contraction semigroup. This proves the theorem.

Note that in general $2\pi\gamma(w)$ defines a tempered but *unbounded* measure, and so the classical Bochner theorem on nonnegative definite functions is not applicable in this case. For this reason the result just proven is a strengthening of Theorem 4.1.

5. Passive systems and dissipative operators. In [1] Dolph studied systems $\mathcal{L}u = \dot{u} - Au = f$, where u, f are L_2 Hilbert space valued functions of a real variable and A is a dissipative operator on H_0. Under suitable conditions on A (see last paragraph below) he showed that a solution to $\mathcal{L}u = f$ exists and satisfies

$$\text{Re} \int_{-\infty}^{t} (f, u)\, dt \geqq 0 \tag{5.1}$$

for all t and all $u \in \mathfrak{D}(\mathcal{L})$; he called $\mathcal{L}u = f$ a passive Hilbert system. The condition (5.1) is the usual requirement in determining when a system described by a linear and time-invariant operator is passive in the sense that the net energy delivered to (absorbed by) the system is nonnegative at any time (see, for example, Wohlers and Beltrami [10]).

Consider now $\mathcal{L}u = f$, $u(0) = 0$, in the case where u, f are L_2 functions which take on values in E^n, and A is also defined on E^n. In this case, it may be shown that the passivity condition (5.1) is equivalent to asserting that $\text{Re}\,(Au, u) \leqq 0$ for all $u \in E^n$, viz., that A is dissipative. In turn, the following condition is also equivalent: the fundamental matrix or semigroup e^{At} satisfies $\| e^{At} \| \leqq 1$. Now consider the system described by the inverse operator $\mathcal{L}^{-1}$. Then $\mathcal{L}^{-1}$ is a linear, time-invariant and, if A is dissipative, passive operator. The Green's kernel W_t of this operator is related to the fundamental matrix by $W_t = e^{At}$ for $t \geqq 0$ with zero elsewhere and the Laplace transform (or immittance) of W_t is the resolvent $(I\lambda - A)^{-1}$. More generally, we have the following theorem due to Zemanian [11].

THEOREM 5.1. *Let W be an operator-valued Schwartz distribution (i.e., with values in $\mathcal{L}(E^n, E^n)$) whose support is the positive t-axis. Then a necessary and sufficient condition in order that W be the kernel of a passive system is that its distributional Laplace transform or immittance be positive real. If W is such a kernel, then it is bounded (Schwartz class D'_{L_∞}).*

Thus we observe to what extent the study of dissipative operators relates to the study of passive systems in the usual sense. It would be instructive to complete the analogy by examining Hilbert space valued systems. For example, one has formally that R_λ is the Laplace transform of the operator-valued (i.e., with values in $\mathcal{L}(H_0, H_0)$) distribution $W(t)$ defined by $W(t) = S(t)$, $t \geqq 0$, and zero elsewhere. We then invert the identity $(I\lambda - A)R_\lambda = I$ to obtain, distributionally, that $\dot{W} - AW = I\delta$, where δ is the Dirac measure. Thus W is the Green's kernel or fundamental solution for the passive Hilbert system under the single assumption that A generates a tempered semigroup $S(t)$ for, in this case, R_λ will be holomorphic in $\text{Re}\,\lambda > 0$ and the inversion is valid.

The program now is to establish a version of Theorem 5.1 in the setting of Hilbert space valued inputs f and outputs u. Presumably this would extend the scope of Dolph's investigations on passive Hilbert systems. The appropriate tool here is the theory of vector-valued distributions (see, for example, [12]).

Finally, we remark that a closed linear operator A having dense domain defines a *passive* Hilbert system $\dot{u} - Au = f$ whenever A is maximal dissipative; this is the "suitable condition" of Dolph referred to above. The significance of Theorem 4.2 in conjunction with this remark is that if the evolution of the Hilbert system is determined by a semigroup $S(t)$, $t \geqq 0$, and if A generates $S(t)$, then the system is passive under the sole condition that the resolvent of A is positive real.

REFERENCES

[1] C. L. Dolph, *Positive real resolvents and linear passive Hilbert systems*, Ann. Acad. Sci. Fenn. Ser. A I, 336/9 (1963), pp. 331–339.

[2] E. Hille and R. S. Phillips, *Functional Analysis and Semi-groups*, Colloquium Publications, vol. 31, 2nd ed., American Mathematical Society, Providence, 1957.

[3] R. S. Phillips, *Dissipative operators and hyperbolic systems*, Trans. Amer. Math. Soc., 90 (1959), pp. 193–254.

[4] W. Cauer, *The Poisson integral for functions with positive real parts*, Bull. Amer. Math. Soc., 38 (1932), pp. 713–715.

[5] J. A. Shohat and J. D. Tamarkin, *The Problem of Moments*, Mathematical Surveys, vol. I, American Mathematical Society, New York, 1943.

[6] E. J. Beltrami and M. R. Wohlers, *Distributions and the Boundary Values of Analytic Functions*, Academic Press, New York, 1966.

[7] L. Schwartz, *Théorie des Distributions*, vols. I and II, Hermann, Paris, 1957 and 1959.

[8] F. Riesz and B. Sz. Nagy, *Functional Analysis*, Ungar, New York, 1955.

[9] M. Schreiber, *Absolutely continuous operators*, Duke Math. J., 29 (1962), pp. 175–190.

[10] M. R. Wohlers and E. J. Beltrami, *Distribution theory as the basis of generalized passive network analysis*, IEEE Trans. Circuit Theory, CT-12 (1965), pp. 164–169.

[11] A. Zemanian, *An N-port realizability theory based on the theory of distributions*, Ibid., CT-10 (1963), pp. 265–274.

[12] L. Schwartz, *Théorie des distributions à valeurs vectorielles*, Ann. Inst. Fourier (Grenoble), 7 (1957), pp. 1–141.

ON LINEAR PASSIVE n-PORTS WITH TIME-VARYING ELEMENTS*

VACLAV DOLEZAL†

Summary. This paper presents a theorem stating conditions under which an n-port obtained by series and parallel connections of elementary n-ports is passive, i.e., incapable of producing energy under any excitation. An electrical n-port built up from purely inductive, resistive and capacitive time-varying n-ports by combining them in series and parallel is a type of the n-port considered.

1. Let $\mathfrak{D}'$ be the set of all n-vector distributions each of which has a finite order and vanishes on the set $(-\infty, 0)$. Stated more explicitly, $f \in \mathfrak{D}'$ if and only if there exists an integer $r \geqq 0$ and a locally integrable n-vector function $F(t)$ vanishing almost everywhere in $(-\infty, 0)$ such that $f = F^{(r)}$, i.e.,

$$\langle f, \phi \rangle = (-1)^r \int_{-\infty}^{\infty} F(t)\phi^{(r)}(t)\, dt \tag{1}$$

for every (one-dimensional) testing function $\phi(t)$ with compact support.

Observe that, with a particular r, the vector function $F(t)$ vanishing on $(-\infty, 0)$ is determined uniquely up to a set of measure zero; thus, let $F(t)$ be called the generating vector function for $f \in \mathfrak{D}'$.

The set $\mathfrak{D}'$ is clearly a linear space.

Let $f \in \mathfrak{D}'$, $f = F^{(r)}$, and let k be a positive integer; if $1 \leqq k \leqq r$, put $f^{(-k)} = F^{(r-k)}$ (distributional derivative); if $k > r$ let $f^{(-k)}$ be the vector distribution corresponding to the locally integrable vector function $\int_0^t [(t-\tau)^{k-r-1}/(k-r-1)!]F(\tau)\, d\tau$, $t \in (-\infty, \infty)$.

It is clear that $f^{(-k)} \in \mathfrak{D}'$; moreover, $f^{(-k)}$ is defined uniquely, i.e., $f^{(-k)}$ does not depend on the choice of r. Actually, if $f = F^{(r)}$, then also $f = G^{(r+1)}$ with $G(t) = \int_0^t F(\tau)\, d\tau$. Then we have, by definition, $f^{(-k)} = G^{(r+1-k)} = (G')^{(r-k)} = F^{(r-k)}$ provided $r - k \geqq 0$; the uniqueness for $r - k < 0$ follows in the same manner.

* Received by the editors October 13, 1966, and in revised form November 29, 1966. Selected by the editors to be included in this collection of papers presented at the Symposium on "The Applications of Generalized Functions" sponsored by the Air Force Office of Scientific Research at the 1966 Fall Meeting of Society for Industrial and Applied Mathematics held at the State University of New York at Stony Brook, September 12–14, 1966.

† Czechoslovakian Academy of Sciences, Matematický Ustav, Praha 1-Nové Město, Zitná 25, Czechoslovakia. This work was supported by the Air Force Cambridge Research Laboratories under Contract AF 19(628)-2981 while the author was at the State University of New York at Stony Brook, Stony Brook, Long Island, New York.

From the above definition we immediately have the following assertion.

If $f \in \mathfrak{D}'$ and k, m are integers, then

$$(f^{(k)})^{(m)} = f^{(k+m)}. \tag{2}$$

(If $k > 0, f^{(k)}$ denotes the kth distributional derivative of f; for $k = 0$ we set $f^{(0)} = f$.)

Let $a(t)$ be an $n \times n$ matrix function defined on $[0, \infty)$; $a(t)$ will be called smooth if $a(t)$ possesses all (ordinary) derivatives on $[0, \infty)$. (At $t = 0$ we understand derivatives from the right.)

Let $a(t)$ be a smooth matrix function, $f \in \mathfrak{D}'$, $f = F^{(r)}$, where $F(t)$ is a generating function for f; then let

$$af = \sum_{i=0}^{r} (-1)^i \binom{r}{i} (a^{(i)}F)^{(r-i)}, \tag{3}$$

and call af the product of a with f.

Note. In (3), $a^{(i)}F$ signifies the product of matrix $a^{(i)}$ with the vector F which vanishes for $t < 0$; clearly, $a^{(i)}F$ is a locally integrable vector function so that the distributional derivative $(a^{(i)}F)^{(r-i)}$ belongs to $\mathfrak{D}'$. Hence, $af \in \mathfrak{D}'$.

It can be easily verified that the definition (3) is meaningful, i. e., af is independent of the choice of r. Moreover, employing the same method of proof as that used in [1, p. 116] for $n = 1$, we can easily prove the formulas:

$$a(bf) = (ab)f, \tag{4}$$

$$(af)' = a'f + af', \tag{5}$$

$$af^{(k)} = \sum_{i=0}^{k} (-1)^i \binom{k}{i} (a^{(i)}f)^{(k-i)}. \tag{6}$$

(Here, $f \in \mathfrak{D}'$, a, b are smooth matrix functions and k is a positive integer.)

Next, let us introduce a certain product of $f \in \mathfrak{D}'$ with a matrix in two variables, which is a generalization of the Volterra integral.

Let $W(t, \tau)$ be an $n \times n$ matrix function defined on the region $R = \{(t, \tau): 0 \leqq \tau \leqq t < \infty\}$; the matrix $W(t, \tau)$ will be called smooth if $W(t, \tau)$ possesses all partial derivatives on R. (On the boundaries $\tau = 0$ and $t = \tau$ of R we understand derivatives from the inside of R.)

If $H(t, \tau)$ is smooth, let $H^* = H(t, t)$; evidently, H^* is a smooth matrix in one variable t.

Let $f \in \mathfrak{D}'$, $f = F^{(r)}$, F being the generating vector function, and let $W(t, \tau)$ be a smooth matrix function; then put

$$[Wf] = \sum_{i=0}^{r-1} (-1)^i \left(\frac{\partial^i W}{\partial \tau^i}\right)^* F^{(r-i-1)} + (-1)^r \int_0^t \frac{\partial^r W(t, \tau)}{\partial \tau^r} F(\tau)\, d\tau. \tag{7}$$

Note. Because $F^{(r-i-1)} \in \mathfrak{D}'$ and $(\partial^i W/\partial\tau^i)^*$ is smooth, the product $(\partial^i W/\partial\tau^i)^* F^{(r-i-1)}$ is well defined and belongs to $\mathfrak{D}'$. Similarly, the vector function $\int_0^t (\partial^r W(t, \tau)/\partial\tau^r)F(\tau)\,d\tau$ is continuous and vanishes for $t < 0$, i. e., the corresponding vector distribution appearing in (7) belongs to $\mathfrak{D}'$ too. Hence, $[Wf] \in \mathfrak{D}'$.

Observe that if particularly $r = 0$ (i. e., f is a regular vector distribution), then $f = F$ and (7) yields $[Wf] = \int_0^t W(t, \tau)F(\tau)\,d\tau$; hence, $[Wf]$ is a generalization of the Volterra integral.

Using the same method as in [1] for $n = 1$, we can easily verify that $[Wf]$ is defined uniquely by (7), i. e., $[Wf]$ is independent of the choice of r.

From the above definition it follows immediately that the product $[Wf]$ is bilinear in W and f.

Combining the product (3) with (7), we can easily prove the following assertion.

Let $a(t)$ be a smooth matrix function in one variable, $W(t, \tau)$ a smooth matrix function in two variables and let $f \in \mathfrak{D}'$; then we have

$$a[Wf] = [(aW)f], \tag{8}$$

$$[W(af)] = [(Wa_\tau)f], \tag{9}$$

where $aW = a(t)W(t, \tau)$ and $Wa_\tau = W(t, \tau)a(\tau)$.

Furthermore, immediately from definition (7) we obtain a second assertion.

If $W(t, \tau)$ is smooth, $f \in \mathfrak{D}'$ and k is a nonnegative integer, then

$$[Wf^{(k)}] = \sum_{i=0}^{k-1} (-1)^i \left(\frac{\partial^i W}{\partial\tau^i}\right)^* f^{(k-i-1)} + (-1)^k \left[\frac{\partial^k W}{\partial\tau^k} f\right], \tag{10}$$

$$[Wf]^{(k)} = \sum_{i=0}^{k-1} \left\{\left(\frac{\partial^i W}{\partial t^i}\right)^* f\right\}^{(k-i-1)} + \left[\frac{\partial^k W}{\partial t^k} f\right]. \tag{11}$$

(The proof of (8) through (11) is formally the same as that for $n = 1$, which is given in [1].)

Let I be the unit $n \times n$ matrix, and for $k = 1, 2, 3, \cdots$, put

$$u_k(t, \tau) = \frac{(t - \tau)^{k-1}}{(k - 1)!} I; \tag{12}$$

then we have for any $f \in \mathfrak{D}'$,

$$f^{(-k)} = [u_k f]. \tag{13}$$

(The proof follows easily from (7), (12) and the definition of $f^{(-k)}$.)

Let us now consider the iterated product (7). To this purpose, let us introduce the following notation.

If $W_1(t, \tau)$ and $W_2(t, \tau)$ are continuous matrix functions, put

$$(14) \qquad (W_1 \times W_2)(t, \tau) = \int_\tau^t W_1(t, \xi)W_2(\xi, \tau)\, d\xi$$

for every $(t, \tau) \in R$.

From (14) it is clear that $W_1 \times W_2$ is again a continuous matrix function; moreover, the product $W_1 \times W_2$ is bilinear in W_1 and W_2.

Using the Fubini theorem, we can immediately prove the following assertion.

If W_1, W_2, W_3 are continuous matrix functions, then

$$(15) \qquad W_1 \times (W_2 \times W_3) = (W_1 \times W_2) \times W_3 .$$

Due to this fact we can define a "power" by $W^{\times k} = W^{\times(k-1)} \times W$, $k = 2, 3, 4, \cdots$, $W^{\times 1} = W$. Then we have that, for any positive integers k, m,

$$(16) \qquad W^{\times k} \times W^{\times m} = W^{\times(k+m)}.$$

Now, we have the next assertion.

Let $W_1(t, \tau)$ and $W_2(t, \tau)$ be smooth matrix functions; then for any $f \in \mathfrak{D}'$ we have

$$(17) \qquad [W_1[W_2 f]] = [(W_1 \times W_2)f].$$

Let us prove this statement by induction. First it can be easily verified that (17) is true for any regular vector distribution f (i. e., for f with $r = 0$), because in this case (17) follows immediately from the Fubini theorem. Thus, assume that (17) holds for any smooth matrix functions W_1, W_2 and any $f \in \mathfrak{D}'$ such that $f = F^{(k)}$ with $0 \leqq k \leqq r - 1$. If $g \in \mathfrak{D}'$ is such that $g = G^{(r)}$, let $y = g^{(-1)}$; then we have by the induction assumption and (10),

$$\begin{aligned}
[W_1[W_2\, g]] = [W_1[W_2\, y']] &= \left[W_1\left(W_2{}^*y - \left[\frac{\partial W_2}{\partial \tau}\, y\right]\right)\right] \\
&= [W_1(W_2{}^*y)] - \left[W_1\left[\frac{\partial W_2}{\partial \tau}\, y\right]\right] \\
&= [(W_1\, W_{2\tau}^*)y] - \left[\left(W_1 \times \frac{\partial W_2}{\partial \tau}\right)y\right] \\
&= \left[\left(W_1\, W_{2\tau}^* - W_1 \times \frac{\partial W_2}{\partial \tau}\right)y\right] = -\left[\left(\frac{\partial}{\partial \tau}\,(W_1 \times W_2)\right)y\right].
\end{aligned}$$

On the other hand, again by (10),

$$[(W_1 \times W_2)g] = (W_1 \times W_2)^* y - \left[\left(\frac{\partial}{\partial \tau}(W_1 \times W_2)\right) y\right].$$

Hence, because $(W_1 \times W_2)^* \equiv 0$,

$$[W_1[W_2 g]] = [(W_1 \times W_2)g],$$

i. e., (17) is true for any vector distribution "of order r." This completes the proof.

Let us now turn to the kernel of our considerations; we are going to introduce a class of operators on $\mathfrak{D}'$ which permits us to prove the main theorem on passive n-ports.

Let σ be the set of all operators defined on $\mathfrak{D}'$ which have the following properties. If $A \in \sigma$ then there exist a smooth real matrix function $W(t, \tau)$ and integers $k \geqq 0$, $g > 0$ with $|k - g| \leqq 1$ such that:

(a) $(\partial^i W/\partial t^i)^* \equiv 0$ for $i = 0, 1, \cdots, g-2$, $(\partial^{g-1} W/\partial t^{g-1})^*$ is a positive definite (not necessarily symmetric) matrix for every $t \geqq 0$;

(b) $Ax = [Wx]^{(k)}$ for every $x \in \mathfrak{D}'$.

It is clear that any $A \in \sigma$ is a linear operator which maps $\mathfrak{D}'$ into itself.

LEMMA 1. *An operator A on $\mathfrak{D}'$ belongs to σ if and only if*

$$Ax = a_1 x' + a_0 x + [\tilde{W} x] \tag{18}$$

for every $x \in \mathfrak{D}'$, where a_1, a_0, $\tilde{W}$ are smooth real matrix functions, and any one of conditions C_1, C_0, C_{-1} is satisfied:

C_1: *a_1 is positive definite for every $t \geqq 0$;*

C_0: *$a_1 \equiv 0$ and a_0 is positive definite for every $t \geqq 0$;*

C_{-1}: *$\tilde{W}^*$ is positive definite for every $t \geqq 0$ and $a_1 \equiv 0$, $a_0 \equiv 0$.*

Proof. First, let $A \in \sigma$. By (b) and (11) we have

$$Ax = \sum_{i=0}^{k-1} \left\{ \left(\frac{\partial^i W}{\partial t^i} \right)^* x \right\}^{(k-i-1)} + \left[\frac{\partial^k W}{\partial t^k} x \right]. \tag{19}$$

Now, if $k - g = 1$, we obtain from (19) by (a),

$$Ax = \left\{ \left(\frac{\partial^{g-1} W}{\partial t^{g-1}} \right)^* x \right\}' + \left(\frac{\partial^g W}{\partial t^g} \right)^* x + \left[\frac{\partial^{g+1} W}{\partial t^{g+1}} x \right] \tag{20}$$

for every $x \in \mathfrak{D}'$. Using (5) it follows that Ax has the form (18) with $a_1 = (\partial^{g-1} W/\partial t^{g-1})^*$; hence, C_1 is satisfied.

If $k - g = 0$, then (19) yields

$$Ax = \left(\frac{\partial^{g-1} W}{\partial t^{g-1}} \right)^* x + \left[\frac{\partial^g W}{\partial t^g} x \right] \tag{21}$$

for all $x \in \mathfrak{D}'$, i. e., C_0 holds.

Finally, if $k - g = -1$, then

$$(22) \qquad Ax = \left[\frac{\partial^{g-1}W}{\partial t^{g-1}} x\right]$$

for all $x \in \mathfrak{D}'$, and C_{-1} holds.

Conversely, let the operator A be defined on $\mathfrak{D}'$ by (18), and let C_1 hold. Using (5), (2), (9), (13) and (17), we can write for every $x \in \mathfrak{D}'$,

$$\begin{aligned} Ax &= (a_1x)' + (a_0 - a_1')x + [\tilde{W}x] = \{(a_1x)^{(-1)} + (b_0x)^{(-2)} + [\tilde{W}x]^{(-2)}\}^{(2)} \\ &= \{[u_1(a_1x)] + [u_2(b_0x)] + [u_2[\tilde{W}x]]\}^{(2)} \\ &= \{[(u_1a_{1_\tau})x] + [(u_2b_{0_\tau})x] + [(u_2 \times \tilde{W})x]\}^{(2)} = [Wx]^{(2)}, \end{aligned}$$

where we have set $b_0 = a_0 - a_1'$ and

$$(23) \qquad W(t,\tau) = u_1a_1(\tau) + u_2b_0(\tau) + (u_2 \times \tilde{W})(t,\tau).$$

However, $W^* = a_1(t)$, i. e., condition (a) is satisfied with $g = 1$ and also $|k - g| \leqq 1$ is true with the above value $k = 2$; hence, $A \in \sigma$.

The same argument applies if A is given by (18) and either C_0 or C_{-1} hold. This completes the proof.

Using the lemma just proven we can easily prove the following assertion.

LEMMA 2. (a) *If $A, B \in \sigma$, then $A + B \in \sigma$.*

(b) *If $A \in \sigma$, then A is invertible, i. e., the inverse operator A^{-1} exists and we have $A^{-1} \in \sigma$.*

Proof. The assertion (a) is trivial; for its proof, it suffices to realize that A and B can be represented by (18) and the fact that the sum of two positive definite matrices is again positive definite. Thus, let us turn to the proof of (b). If $A \in \sigma$, then by condition (a) and (11) we can write for every $x \in \mathfrak{D}'$,

$$(24) \qquad \begin{aligned} Ax = [Wx]^{(k)} = \{[Wx]^{(g)}\}^{(k-g)} &= \left\{\left(\frac{\partial^{g-1}W}{\partial t^{g-1}}\right)^* x + \left[\frac{\partial^{g}W}{\partial t^{g}} x\right]\right\}^{(k-g)} \\ &= \{a(x + [\hat{W}x])\}^{(k-g)}, \end{aligned}$$

where we have set $a = (\partial^{g-1}W/\partial t^{g-1})^*$, which is positive definite by assumption, and $\hat{W}(t, \tau) = a^{-1}(t)\partial^g W(t, \tau)/\partial t^g$. (Note that $a^{-1}(t)$ exists and is also positive definite for all $t \geqq 0$.)

Next, define the matrix function $H(t, \tau)$ by

$$(25) \qquad H(t, \tau) = \sum_{i=1}^{\infty} (-1)^i \hat{W}^{\times i}.$$

We are going to show that the series in (25) converges uniformly on every bounded region $R_T = \{(t, \tau): 0 \leqq \tau \leqq t \leqq T\}$, and consequently, $H(t, \tau)$ is a continuous matrix function on R. Choosing a $T > 0$, put $M = \sup_{(t,\tau) \in R_T} \|\hat{W}(t, \tau)\|$, where $\|\cdot\|$ signifies a matrix norm.

Using the induction principle we can easily verify that for any $(t, \tau) \in R_T$ we have

$$\|\hat{W}^{\times i}(t, \tau)\| \leqq M^i |t - \tau|^{i-1}/(i-1)!; \tag{26}$$

thus, for any $(t, \tau) \in R_T$ the terms of series (25) are majorized in norm by terms of a convergent series $\sum_{i=1}^{\infty} M^i T^{i-1}/(i-1)!$, and consequently, (25) converges uniformly on R_T.

Using the uniform convergence of (25), we can write for any $(t, \tau) \in R$,

$$\begin{aligned}(\hat{W} \times H)(t, \tau) &= \int_\tau^t \left(\hat{W}(t, \xi) \sum_{i=1}^{\infty} (-1)^i \hat{W}^{\times i}(\xi, \tau)\right) d\xi \\ &= \sum_{i=1}^{\infty} \int_\tau^t (-1)^i \hat{W}(t, \xi) \hat{W}^{\times i}(\xi, \tau)\, d\xi \\ &= \sum_{i=1}^{\infty} (-1)^i \hat{W}^{\times(i+1)} = -\hat{W} - H.\end{aligned} \tag{27}$$

Hence, for any $(t, \tau) \in R$ we have the equality

$$\hat{W} + H + \hat{W} \times H = 0. \tag{28}$$

By the same argument it follows that

$$\hat{W} + H + H \times \hat{W} = 0 \tag{29}$$

on R.

Furthermore, using both (28) and (29) and the induction principle, we can easily verify that $H(t, \tau)$ possesses all partial derivatives on R, i. e., $H(t, \tau)$ is a smooth matrix function.

Now, define the operator B on $\mathfrak{D}'$ by

$$Bx = a^{-1} x^{(g-k)} + [(H a_\tau^{-1}) x^{(g-k)}]. \tag{30}$$

(Note that a^{-1} is a smooth matrix function.) Then for any $x \in \mathfrak{D}'$ we have by (24),

$$\begin{aligned}(BA)x = B(Ax) &= a^{-1}(Ax)^{(g-k)} + [(H a_\tau^{-1})(Ax)^{(g-k)}] \\ &= x + [\hat{W}x] + [H(x + [\hat{W}x])] \\ &= x + [(\hat{W} + H + H \times \hat{W})x] = x\end{aligned}$$

by (29).

Using the same computation and (28) we conclude that $(AB)x = x$ for any $x \in \mathfrak{D}'$; hence, A is invertible and $B = A^{-1}$.

It remains to show that actually $A^{-1} \in \sigma$. If $k - g = -1$, then (30) yields by (10) with $\hat{H} = H a_\tau^{-1}$,

$$A^{-1}x = a^{-1}x' + \hat{H}^*x - \left[\frac{\partial \hat{H}}{\partial \tau} x\right];$$

consequently, by Lemma 1, $A^{-1} \in \sigma$.

The case $k - g = 0$ is obvious. Finally, if $k - g = 1$, we have by (30) and the above formulas, $A^{-1}x = a^{-1}x^{(-1)} + [\hat{H}x^{(-1)}] = a^{-1}[u_1x] + [\hat{H}[u_1x]] = [\tilde{W}x]$ with $\tilde{W}(t, \tau) = a^{-1}(t)u_1 + (\hat{H} \times u_1)(t, \tau)$. However, $\tilde{W}^* = a^{-1}(t)$, and consequently, $\tilde{g} = 1$, $\tilde{k} = 0$ by notation used in the definition of σ. Hence, $A^{-1} \in \sigma$ and Lemma 2 is proven.

Now, we are going to consider a subset of σ; to this purpose, let us introduce the following notation.

Let $\mathfrak{L}_0 \subset \mathfrak{D}'$ be the set of all regular n-vector distributions, i. e., $f \in \mathfrak{L}_0$ if and only if f is a vector distribution corresponding to a locally integrable vector function. Moreover, let $\mathfrak{L}_1 \subset \mathfrak{L}_0$ be the set of all regular vector distributions corresponding to absolutely continuous vector functions, i. e., $f \in \mathfrak{L}_1$ if and only if $f(t)$ is absolutely continuous on $(-\infty, \infty)$ (and, evidently, vanishes for $t < 0$).

If $x \in \mathfrak{L}_1$ and $y \in \mathfrak{L}_0$, let

$$(x, y)_t = \int_0^t x^T(\tau)y(\tau)\,d\tau \tag{31}$$

for every $t \geqq 0$, where x^T denotes the transpose of the vector x.

Note. Because both x and y in (31) are regular vector distributions, we denote the corresponding generating vector functions by $x(t)$ and $y(t)$, respectively.

Let $p \subset \sigma$ be the set of all operators A which satisfy the condition

$$(x, Ax)_t \geqq 0 \tag{32}$$

for every $x \in \mathfrak{L}_1$ and every $t \geqq 0$.

Observe that this definition is meaningful, because by Lemma 1 we have for any $A \in \sigma, x \in \mathfrak{D}', Ax = a_1x' + a_0x + [\tilde{W}x]$ (where, eventually, $a_1 = 0$ or $a_1 = a_0 = 0$) and consequently, $Ax \in \mathfrak{L}_0$ for any $x \in \mathfrak{L}_1$.

Now we can state the main assertion.

LEMMA 3. *If $A, B \in p$, then $A + B \in p$ and $A^{-1} \in p$.*

Proof. The first assertion is trivial; if $A, B \in p$, then $A, B \in \sigma$, and consequently, $A + B \in \sigma$ by Lemma 2. On the other hand, for any $x \in \mathfrak{L}_1$ and $t \geqq 0$, we have $(x, (A + B)x) = (x, Ax)_t + (x, Bx)_t \geqq 0$. Hence, $A + B \in p$.

In order to prove the second assertion, let us recall the definition of the set σ, and for every $A \in \sigma$ define the number $r(A)$ as $k - g$, where integers k, g are defined in conditions (a), (b). Observe that $r(A)$ is determined uniquely by A, i.e., it does not depend on the choice of k. Then from the proof of Lemma 2, particularly from (30), it follows that $r(A^{-1}) = -r(A)$ for any $A \in \sigma$.

Thus, let $A \in p$; then $A \in \sigma$, and consequently, A^{-1} exists and belongs to σ. Choosing an $x \in \mathfrak{L}_1$ and $t \geqq 0$, consider the value $\mathfrak{J} = (x, A^{-1}x)_t$. Denoting $A^{-1}x = y$, we have $x = Ay$, and consequently,

$$\mathcal{J} = (Ay, y)_t = (y, Ay)_t = \int_0^t y^T(\tau)(Ay)(\tau)\, d\tau. \tag{33}$$

Referring to Lemma 1, let us distinguish three cases according to condition C_i which is satisfied by A.

Case 1. If A satisfies condition C_1, then $r(A) = 1$, and consequently, $r(A^{-1}) = -1$, i.e., the operator A^{-1} satisfies condition C_{-1}. Thus, by (18), $y = [\tilde{W}x]$ so that for the vector function $y(t)$ corresponding to y we have $y(t) = \int_0^t W(t, \tau)x(\tau)\, d\tau$. However, from this it follows that $y \in \mathfrak{L}_1$, because we have $y(t) = \int_0^t u(\tau)\, d\tau$ with $u(t) = \tilde{W}^* x(t) + \int_0^t \frac{\partial \hat{W}}{\partial t} x(\tau)\, d\tau$. Hence, by (33) we have $\mathfrak{J} \geqq 0$ for every $t \geqq 0$.

Case 2. If A satisfies condition C_0, then A^{-1} also satisfies C_0 and we have by (18), $y = a_0 x + [\hat{W}x]$; because $x \in \mathfrak{L}_1$, we have, by the same argument as before, $y \in \mathfrak{L}_1$. Hence, by (33), $\mathfrak{J} \geqq 0$, for every $t \geqq 0$.

Case 3. Finally, let A satisfy condition C_{-1}; then A^{-1} satisfies C_1, so that $y = a_1 x' + a_0 x + [\hat{W}x]$. Consequently, we have only $y \in \mathfrak{L}_0$, i.e., we cannot use directly (32) for $(y, Ay)_t$ in (33). However, we can easily circumvent this difficulty as follows: Choose a $t^* > 0$ and keep it fixed in the next consideration. Choosing an arbitary $\epsilon > 0$ we can find a vector function $\tilde{z}(t)$ which is absolutely continuous on $[0, t^*]$ and such that $\tilde{z}(0) = 0$, and $\int_0^{t^*} \| y(\tau) - \tilde{z}(\tau) \|\, d\tau < \epsilon$. If we define $z(t) = \tilde{z}(t)$ on $[0, t^*]$, $z(t) = 0$ for $t < 0$ and $z(t) = \tilde{z}(t^*)$ for $t > t^*$, then we evidently have $z \in \mathfrak{L}_1$.

Next, we can write

$$\begin{aligned} \lambda &= (y, Ay)_{t^*} - (z, Az)_{t^*} \\ &= \int_0^{t^*} y^T(\tau)A(y - z)(\tau)\, d\tau + \int_0^{t^*} (y(\tau) - z(\tau))^T A(z - y)(\tau)\, d\tau \\ &\quad + \int_0^{t^*} (y(\tau) - z(\tau))^T Ay(\tau)\, d\tau. \end{aligned} \tag{34}$$

Consequently,

$$|\lambda| \leqq \int_0^{t^*} \| y(\tau) \| \cdot \| A(y - z) \|\, d\tau$$

$$(35) \qquad + \int_0^{t^*} \| y(\tau) - z(\tau) \| \cdot \| A(z - y) \| \, d\tau$$
$$+ \int_0^{t^*} \| y(\tau) - z(\tau) \| \cdot \| Ay \| \, d\tau.$$

On the other hand, because A satisfies the condition C_{-1}, we have $Au = \int_0^t Q(t, \tau) u(\tau) \, d\tau$ for any $u \in \mathfrak{L}_1$, where $Q(t, \tau)$ is a smooth matrix, and consequently, for any $t \in [0, t^*]$,

$$(36) \quad \| (Au)(t) \| \leqq \int_0^t \| Q(t, \tau) \| \cdot \| u(\tau) \| \, d\tau \leqq \mathfrak{K} \int_0^{t^*} \| u(\tau) \| \, d\tau,$$

where we have set $\mathfrak{K} = \sup_{(t,\tau) \in R_{t^*}} \| Q(t, \tau) \|$. Using estimate (36) and the above inequality for $y(t) - z(t)$, we get from (35),

$$(37) \quad \begin{aligned} | \lambda | &< \int_0^{t^*} \| y(\tau) \| \mathfrak{K} \epsilon \, d\tau + \int_0^{t^*} \| y(\tau) - z(\tau) \| \mathfrak{K} \epsilon \, d\tau \\ &\quad + \int_0^{t^*} \| y(\tau) - z(\tau) \| \mathfrak{K} \int_0^{t^*} \| y(\omega) \| \, d\omega \, d\tau \\ &< 2M\mathfrak{K}\epsilon + \mathfrak{K}\epsilon^2, \end{aligned}$$

where we denote $M = \int_0^{t^*} \| y(\tau) \| \, d\tau$.

However, because $z \in \mathfrak{L}_1$, we have by assumption (32), $(z, Az)_{t^*} \geqq 0$; moreover, since $\epsilon > 0$ can be chosen as small as we please, we get from (37) and (34), $(y, Ay)_{t^*} \geqq 0$, and consequently, by (33), $\mathfrak{J} \geqq 0$. This completes the proof.

Note. The above development may be simplified if we adopt the following approach. Let C^∞ be the set of all vector distributions corresponding to infinitely differentiable functions with support contained in $[0, \infty)$; if $x, y \in C^\infty$, let $(x, y)_t$ be defined by (31). Furthermore, let $p^* \subset \sigma$ be the set of all operators A which satisfy condition (32) for every $x \in C^\infty$ and every $t \geqq 0$. Then from Lemma 1 it follows that $Ax \in C^\infty$ whenever $x \in C^\infty$ and $A \in p^*$. Using this fact and Lemma 2(b) we immediately conclude that $(x, A^{-1}x)_t \geqq 0$ for every $x \in C^\infty$ and $A \in p^*$, i.e., that Lemma 3 is true with p replaced by p^*. However, using a similar procedure as in the proof of Lemma 3, the reader can easily verify that both systems p and p^* are identical.

In order to simplify the wording of the main theorem, let us introduce the following notation.

Let $a(t)$ be a smooth, real and symmetric matrix function; $a(t)$ will be called "positive" if $a(t)$ is a positive definite matrix for every $t \geqq 0$. Moreover, $a(t)$ will be called "strongly positive" if $a(t)$ is a positive and $a'(t)$ a positive semidefinite matrix for every $t \geqq 0$.

Then we have the following assertion.

LEMMA 4. (a) *If $a(t)$ is positive, then the operator A on $\mathfrak{D}'$ defined by $Ax = ax$ belongs to p.*

(b) *If $a(t)$ is strongly positive, then the operator B on $\mathfrak{D}'$ defined by $Bx = (ax)'$ belongs to p.*

Proof. (a) is trivial; thus we prove (b) only. Because $Bx = ax' + a'x$ for every $x \in \mathfrak{D}'$ by (5), we have $B \in \sigma$ by Lemma 1. Furthermore, let $x \in \mathfrak{L}_1$ and $t \geqq 0$; using the fact that $x(0) = 0$, we obtain

$$
\begin{aligned}
(x, Bx)_t &= \int_0^t x^T(\tau)(a(\tau)x(\tau))'\,d\tau \\
&= x^T(t)a(t)x(t) - \int_0^t x^{T\prime}(\tau)a(\tau)x(\tau)\,d\tau.
\end{aligned} \tag{38}
$$

On the other hand,

$$
\begin{aligned}
&\int_0^t x^T(\tau)(a(\tau)x(\tau))'\,d\tau \\
&\qquad = \int_0^t x^T(\tau)a'(\tau)x(\tau)\,d\tau + \int_0^t x^T(\tau)a(\tau)x'(\tau)\,d\tau.
\end{aligned} \tag{39}
$$

Combining (38) and (39) and using the symmetry of $a(t)$, we obtain immediately,

$$
(x, Bx)_t = \frac{1}{2}x^T(t)a(t)x(t) + \frac{1}{2}\int_0^t x^T(\tau)a'(\tau)x(\tau)\,d\tau \geqq 0;
$$

hence, $B \in p$ and (b) is proven.

2. Let us now turn our attention to the physical interpretation of the above results.

A (linear) n-port N will be said to have the admittance A (A is a linear operator from $\mathfrak{D}'$ into itself), if the signal $s \in \mathfrak{D}'$ and the response $r \in \mathfrak{D}'$ of N are related by $r = As$; similarly, N will be said to have the impedance Z if s and r are related by $s = Zr$.

If N_1 and N_2 are n-ports which have admittances A_1 and A_2, respectively, then the parallel combination of N_1 and N_2 is an n-port having the admittance $A_1 + A_2$; analogously, if N_1 and N_2 have impedances Z_1 and Z_2, respectively, then the series combination of N_1 and N_2 is an n-port with the impedance $Z_1 + Z_2$.

Note. We are assuming here that the conditions guaranteeing the validity of the above relations are satisfied; as known, in the case of electrical n-ports these conditions are met, if, for example, each port is equipped with an ideal transformer.

Let D be the operator on $\mathfrak{D}'$ defined by $Dx = x'$; then we can easily state the main theorem.

THEOREM. *Let N be an n-port which can be considered as a result of series and parallel combinations of n-ports $N_1, N_2, \cdots, N_k$. Let each n-port N_i possess the admittance or impedance, which is equal either to Q_i with Q_i being positive, or to DQ_i with Q_i strongly positive. Then N possesses both the admittance A and impedance Z, and we have $A, Z \subset p$.*

The proof follows immediately from Lemmas 3 and 4, the above rules concerning the series and parallel combinations and from the fact that if T is the admittance (impedance) of an n-port K and T is invertible, then T^{-1} is the impedance (admittance) of K.

Observe that the product (32) has a physical meaning of energy supplied into an n-port with admittance A by a signal $x \in \mathfrak{L}_1$ in the time span $[0, t]$; hence, any n-port satisfying the hypothesis of the theorem is passive.

An example of an n-port considered in the theorem is furnished by an electrical n-port, which appears as a series and parallel combination of resistive n-ports characterized by positive matrices $R(t)$, of inductive n-ports characterized by strongly positive matrices $L(t)$, and capacitive n-ports with strongly positive matrices $C(t)$.

REFERENCE

[1] V. DOLEZAL, *Dynamics of Linear Systems*, Publishing House of the Czechoslovak Academy of Sciences, Prague, 1964.

UNRENORMALIZABLE FIELD THEORIES IN TERMS OF LORENTZ-INVARIANT GENERALIZED FUNCTIONS*

W. GÜTTINGER AND E. PFAFFELHUBER†

Abstract, A new approach to unrenormalizable quantum field theories in terms of Lorentz-invariant generalized functions is developed. The techniques of Lorentz-invariant generalized functions are used to discuss the local structure of those theories and to derive perturbative and nonperturbative solutions to singular propagator and Bethe-Salpeter equations of four-fermion theories in Minkowski and Euclidean space. An unsubtracted Lehmann representation for increasing spectral functions is discussed and some problems related to the application of distributions, of their products and Hilbert transforms to quantum field theory are briefly analyzed.

1. Introduction. The aim of the present paper is to discuss some problems relating to the application of the theory of Lorentz-invariant generalized functions to unrenormalizable quantum field theories. Some of the generalized function and field theory concepts used in what follows have already been explained in a previous paper [1]. For further details we refer to a forthcoming paper [2] and to [3]. We shall adopt Schweber's relativistic notation [4]. In particular, a point in Minkowski space R^4 is denoted by x, $x = (x_\nu) = (x_0, \mathbf{x})$, with the metric $x^2 = x_0^2 - \mathbf{x}^2$. The four-momentum is denoted by p, $p = (p_\nu) = (p_0, \mathbf{p})$, with $p^2 = p_0^2 - \mathbf{p}^2$ and $px = p_0x_0 - \mathbf{px}$. The volume element in R^4 is $d^4x = dx_0\, d\mathbf{x}$; the notation $i0$ means, formally, $i\epsilon$ with $\epsilon \to +0$. $\theta(x_0)$ is the step function, $\theta(x_0) = 1$ for $x_0 > 0$, $\theta(x_0) = 0$ for $x_0 < 0$. The Fourier operator F is given by $\tilde{f} = Ff = \int_{-\infty}^{\infty} \exp(ipx) f\, d^4x$.

According to what has been said in the preceding paper [1], the nontime-ordered (or Wightman) two-point function of an interacting boson field $\phi(x)$, $\Delta_+'(x) = -i\langle 0 \mid \phi(x/2)\phi(-x/2) \mid 0\rangle$, and the corresponding time-ordered function $\Delta_F'(x) = 2i\langle 0 \mid T(\phi(x/2)\phi(-x/2)) \mid 0\rangle$ possess the symbolic spectral representations

$$(1.1)\qquad \begin{aligned} \Delta_+'(x) &= \int_0^\infty \rho(m^2)\Delta_+(x, m^2)\, dm^2, \\ \Delta_F'(x) &= \int_0^\infty \rho(m^2)\Delta_F(x, m^2)\, dm^2 \end{aligned}$$

---

* Received by the editors November 3, 1966, and in revised form March 29, 1967. Extended version of a paper contributed at the Symposium on "The Applications of Generalized Functions" sponsored by the Air Force Office of Scientific Research at the 1966 Fall Meeting of Society for Industrial and Applied Mathematics held at the State University of New York at Stony Brook, September 12–14, 1966.

† Department of Physics, University of Munich, München, Germany. This research was supported by the Deutsche Forschungsgemeinschaft.

in virtue of (i) Lorentz invariance and (ii) positive energy spectrum. Here, Δ_+ and Δ_F are the free field functions

$$\Delta_+(x, m^2) = \frac{-i}{(2\pi)^3}\int_{-\infty}^{\infty} e^{-ipx}\theta(p_0)\theta(p^2)\delta(p^2 - m^2)\, d^4p$$

$$(1.2) \qquad = \frac{-m}{8\pi} H_1^{(1)}(m\sqrt{z})/\sqrt{z}\,|_{z=x^2-ix_00},$$

$$\Delta_F(x, m^2) = \frac{2i}{(2\pi)^4}\int_{-\infty}^{\infty} e^{-ipx}\frac{1}{p^2 - m^2 + i0}\, d^4p = \frac{im}{4\pi} H_1^{(2)}(m\sqrt{\xi})/\sqrt{\xi}\,|_{\xi=x^2-i0},$$

respectively ($H_1^{(\nu)}$ is the Hankel function), m being a mass parameter and $\rho(m^2)$ a spectral function. Similarly, the two-point functions S_+', S_F' of an interacting fermion field $\psi(x)$ admit the representations

$$(1.3) \qquad S'_{+,F}(x) = \int_0^{\infty} [\rho_1(m^2)S_{+,F}(x, m^2) + \rho_2(m^2)\Delta_{+,F}(x, m^2)]\, dm^2$$

with spectral functions $\rho_1(m^2)$, $\rho_2(m^2)$ and $S_{+,F} = -(i\gamma\partial + m)\Delta_{+,F}$.

Following Wightman [5], it was customary to view the quantities Δ_+', Δ_F', S_+', S_F' as tempered distributions. However, as we shall see, unrenormalizable field theories (e.g., four-fermion interactions) are characterized by nontempered distributions and by analytic functionals. Therefore, we shall assume our two-point (and similarly the n-point) functions to be generalized functions f acting on a space $\mathbf{\Phi}$ of test functions ϕ depending on the type of interaction. The value of f at ϕ is denoted by $f\langle\phi\rangle$, symbolically, $f\langle\phi\rangle = \int_{-\infty}^{\infty} f(x)\phi(x)\, d^4x$ or $f\langle\phi\rangle = \int_C dx_0 \int f(x)\phi(x)\, d\mathbf{x}$ with some contour C. According to [1], $i\Delta_+'(x)$ (and, similarly, $iS_+'(x)$) is a positive-definite generalized function, $i\Delta_+'\langle\phi * \hat{\phi}\rangle \geqq 0$. Then the generalized Bochner theorem [1], [3] states that $\rho(m^2) \geqq 0$ is a positive measure (and, similarly, $2m\rho_1(m^2) \geqq \rho_2(m^2) \geqq 0$) and that either (i) $\rho(p^2)$ is tempered as $p^2 \to \infty$, i.e., $\rho(p^2) \in \mathbf{S}'$, so that $\Delta_+', \cdots, S_F'$ are tempered distributions $\in \mathbf{S}'$, or (ii) $\rho(p^2)$ (ρ_1, ρ_2) is not tempered as $p^2 \to \infty$ and grows faster than any power of $|p^2|$, i.e., $\rho(p^2)$ is an element of $\mathbf{D}'$ or $\mathbf{K}'\{M_q\}$ or $\mathbf{S}_\alpha'^{\beta}$, so that $\Delta_+', \cdots, S_F'$ are generalized functions in the spaces $\mathbf{Z}'$ or $\mathbf{Z}'\{M_q\}$ or $\mathbf{S}_\beta'^{\alpha}$, i.e., analytic functionals. This follows at once from Parseval's formula, taken in the sense of generalized functions, viz., $\tilde{f}\langle\psi\rangle = (2\pi)^4 f\langle\hat{\phi}\rangle$, where $Ff = \tilde{f}$, $\psi(p) = F\phi(x)$ and $\hat{\phi}(x) = \phi(-x)$. In fact, taking the Fourier transform of $\Delta_+'(x)$, (1.1), we obtain $F\Delta_+'(x) = \tilde{\Delta}_+'(p) = -2\pi i\theta(p_0)\theta(p^2)\rho(p^2)$, so that

$$\theta(p_0)\theta(p^2)\rho(p^2)\langle\psi(p)\rangle = i(2\pi)^3\Delta_+'(x)\langle\hat{\phi}(x)\rangle.$$

For example, if $\rho(p^2)$ is not tempered but grows exponentially, $\rho(p^2)$ may be viewed as an element of $\mathbf{D}'$ (i.e., $\psi \in \mathbf{D}$). Then $\phi = F^{-1}\psi$ is an entire analytic function $\in \mathbf{Z}$, and $\Delta_+'(x)$ is an analytic functional $\in \mathbf{Z}'$.

Field theories of type (i) with tempered spectral functions are called renormalizable; those of type (ii) with nontempered spectral functions are termed unrenormalizable (the latter term being rather misleading, however). In case (i) (e.g., $\rho(p^2) \approx (p^2)^N$ as $p^2 \to \infty$) the propagator $\Delta_F'(x)$ has at most algebraic singularities on the lightcone $x^2 = 0$, while in the case (ii) (e.g., $\rho(p^2) \sim \exp{(p^2)^\alpha}$) $\Delta_F'(x)$ has an essential singularity on the cone (e.g., $\Delta_F'(x) \sim \exp{(x^2)^{-\alpha/(1-2\alpha)}}$). Indeed, we shall see in the next section that in a local (causal) field theory $\rho(p^2)$ cannot grow as a linear exponential, or faster, as $p^2 \to \infty$.

Independently of whether a field theory is renormalizable or not, the following problems arise:

(i) If $\rho(m^2)$ increases as $m^2 \to \infty$, the Fourier operator F applied to the time-ordered function $\Delta_F'(x)$ in (1.1) does not commute with the m^2-integration. In fact, interchanging formally F and $\int dm^2$ we would arrive at the nonexistent Hilbert transform (Lehmann representation [4])

$$F\Delta_F'(x) = \tilde{\Delta}_F'(p) = 2i \int_0^\infty \frac{\rho(m^2)}{p^2 - m^2 + i0}\, dm^2 \tag{1.4}$$

in virtue of (1.2). However, having given a proper meaning to $\Delta_F'(x)$ in the sense of generalized functions, the Fourier transform $\tilde{\Delta}_F'(p)$ certainly cannot possess the representation (1.4). In field theory one usually goes the other way around: If $\rho(m^2)$ increases, the formal integral (1.4) is given sense by—loosely speaking—differentiating it sufficiently often with respect to p^2 and then reintegrating the expression so obtained to yield a finite integral plus an additive polynomial containing a set of unknown integration constants. For example, if $\rho \sim (m^2)^{n-\epsilon}$ as $m^2 \to \infty$, this procedure yields the expression (principal value integral at $m^2 = M^2$, $M^2 > 0$ arbitrary):

$$\begin{aligned}\tilde{\Delta}_F'(p) = 2i(p^2 - M^2)^n \int_0^\infty \frac{\rho(m^2)}{(m^2 - M^2)^n(p^2 - m^2 + i0)}\, dm^2 \\ + \sum_{\mu=0}^{n-1} c_\mu (p^2 - M^2)^\mu .\end{aligned} \tag{1.5}$$

The arbitrary coefficients c_μ are the familiar "subtraction" or "renormalization" constants (to be fixed by experiment). Upon Fourier transforming (1.5), one arrives at a definition of $\Delta_F'(x)$, (1.1), for increasing spectral functions, containing, however, an arbitrary derivative polynomial

$\sum_0^{n-1} a_\mu \Box^\mu \delta(x)$, $\delta(x) = \delta(x_0)\delta(\mathbf{x})$. The procedure can be justified by appealing to the causality postulate. This postulate implies that the retarded function

$$\Delta_R'(x) = -\theta(x_0)[\Delta_+'(x) - \Delta_+'(-x)]$$

vanishes outside the forward cone $x^2 \geqq 0$, $x_0 \geqq 0$, so that its Fourier transform $\tilde{\Delta}_R'(k)$, $k = p + iq$, is, as a function of k^2, analytic in the k^2-plane cut along the positive real axis. $\tilde{\Delta}_R'$ plays then exactly the role of the indicatrix introduced in the preceding paper. If $\tilde{\Delta}_R'(k)$ is tempered, Cauchy's integral formula can be applied to the function $u(k^2) = \tilde{\Delta}_R'(k) /(k^2 - M^2 - i0)^n$ according to Fig. 1 yielding

$$u(k^2) = \frac{(k^2 - M^2)^n}{2\pi i} \int_0^\infty \frac{g(m^2)}{(m^2 - M^2)^n(m^2 - k^2)} dm^2 + u_0(k^2),$$

$$u_0 = \sum_{\mu=0}^{n-1} \frac{u^{(\mu)}(M^2)}{\mu!} (k^2 - M^2)^\mu$$

(where the term $u_0(k^2)$ originates from the limit $\epsilon \to 0$), taking into account that the retarded and advanced functions $\tilde{\Delta}_R'(p)$ and $\tilde{\Delta}_A'(p) = \tilde{\Delta}_R'(-p)$ are boundary values of $u(k^2)$ as $q = \mathscr{I}mk \to 0$, $\tilde{\Delta}_R'(p) = u(p^2 + ip_0 0)$, $\tilde{\Delta}_A'(p) = u(p^2 - ip_0 0)$, i.e.,

$$g(p^2)\theta(p^2) = \lim_{\epsilon \to 0} [u(p^2 + i\epsilon) - u(p^2 - i\epsilon)] = 2\pi i \rho(p^2)\theta(p^2).$$

This reproduces (1.5) with $\Delta_F' = \Delta_+' + \Delta_A'$, $c_\mu = d^\mu \tilde{\Delta}_F'(p)/d(p^2)^\mu \mu!$. Of course, the naive way of "deriving" (1.5) from (1.4) by differentiation and reintegration does not work if $\tilde{\Delta}_F'$ is not tempered. In that case a refined version of the indicatrix techniques yields a spectral representation containing an arbitrary entire function $u_0(k^2)$. The infinite set of unknown constants involved in this u_0 has often been considered as representing a nonlocal structure of unrenormalizable interactions and has even been taken as an indication that unrenormalizable theories are unphysical. The

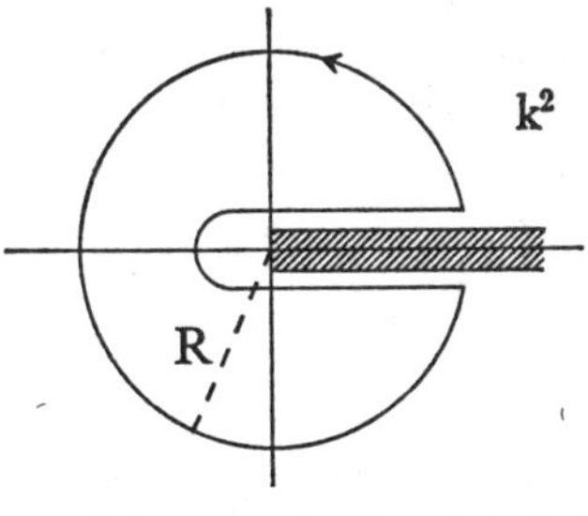

FIG. 1. *The contour for deriving* (1.5) *by Cauchy's integral formula* ($R \to \infty$)

entire arbitrariness involved in this type of formalism originates from the fact that, in giving a meaning to (1.4) in case of an increasing ρ, one appealed solely to the causality postulate and merely exploited the corresponding analyticity properties. Indeed, causality alone is quite a negative statement—telling only what should not happen inside a system—thereby leaving too much space and arbitrariness for the dynamics which are well known to be governed by nonanalytic properties of the system.

In a previous paper [1] we have shown how singular generalized functions such as $x_+^{-\lambda}$, $\lambda > 0$, can be generated without encountering arbitrary "subtraction" constants, except for a single scaling parameter **a** of dimension of length which enters the functionals only if the associated classical integrals exhibit logarithmic divergences. In §3 we extend this procedure to Lorentz-invariant distributions, thereby giving a meaning to the spectral representation (1.1) for $\Delta_F'(x)$. The result is an unsubtracted Lehmann representation in momentum space which does not suffer from the defects of (1.5).

(ii) Closely related to the problem of getting rid of subtraction terms is the question of defining products of generalized functions. This question arises in two ways: (a) Any time-ordered function, such as $\Delta_F'(x)$, is formally (via the causality postulate!) connected with nontime-ordered functions, such as $\Delta_+'(x)$, by multiplying the latter by the step function $\theta(x_0)$. If $\rho(m^2)$ increases (e.g., $\rho \approx m^\alpha$, $\alpha > 0$), $\Delta_+'(x)$ is a singular generalized function (e.g., $\Delta_+'(x) \approx (x^2 - ix_00)^{-2-\alpha/2})$ and the product

$$\Delta_F'(x) = 2i\,[\theta(x_0)\cdot\Delta_+'(x) + \theta(-x_0)\cdot\Delta_+'(-x)]$$

does not exist. So also the Fourier transform $F[\theta(x_0)\cdot\Delta_+'(x)]$, or equivalently (1.4), ceases to make sense. One now either makes the generalized functions into an algebra [3]—which essentially reproduces the subtraction formalism sketched in (i) above—or else one defines products such as $\theta(x_0)\cdot(x^2 - ix_00)^{-\lambda}$ by continuing $\theta(x_0)(x^2 - ix_00)^{-\lambda}$ analytically in λ from $\operatorname{Re}\lambda < 0$ into the complex λ-plane with a suitable interpretation of possibly occurring pole terms. This reproduces the subtraction-free formalism of §3 (cf. also [1], [3]). (b) Products of propagators, $(\Delta_F(x, m^2))^n$, etc., arise from a perturbation formulation of field theories in terms of Feynman diagrams. A definition of these products can be accomplished in either of the two ways indicated above in (a). The familiar divergences of field theoretic perturbation schemes arise from letting such products of generalized functions be undefined.

(iii) Summing formally certain subsets of Feynman diagrams, one arrives at integral equations with singular generalized function kernels, whose formal iterative solutions reproduce the original set of diagrams together with the representative generalized function products. Leaving these undefined, one is left with an infinite series of divergent integrals with the order of divergence increasing with the order of the diagrams in the case of unrenormalizable theories. If, however, one sticks to a proper definition of these integral equations in terms of generalized functions one is able to obtain finite, well-defined solutions which agree with what one obtains by properly defining the generalized function products representing the perturbative diagrams. Equivalently, these integral equations correspond to differential equations with singular generalized functions as coefficients, e.g., $\Box f(x) = -4g(x^2 - i0)^{-3}f(x)$, $\Box = \partial^2/\partial x_0^2 - \Delta$. One can give a meaning to such equations (whose formal solutions, e.g., $f(x) = \exp(\pm\sqrt{g}/(x^2 - i_0))$ (see §5), necessitate a generalized function interpretation) and the solutions obtained coincide with the generalized function solutions of the integral equations. These results, to be discussed in §5, are valid for a certain class of unrenormalizable equations. The resulting perturbative Minkowski space solutions should be confronted with non-perturbative solutions obtainable by defining the singular integral equations for time-ordered functions by the method of complex extension. This problem (related to the so-called peratization procedure) will be analyzed in §6.

2. Local structure of unrenormalizable field theories.

THEOREM. *In a causal (renormalizable or unrenormalizable) field theory, the generalized functions* $\rho(p^2), \cdots, \tilde{\Delta}_F'(p), \cdots, \tilde{S}_F'(p)$, *acting on a space of test functions with compact support (or decreasing at infinity exponentially), satisfy for* $|p^2| \to \infty$ *the high energy bound*

$$(2.1)\qquad |\rho(p^2)| \leqq C\exp(|p^2|^\alpha), \quad \cdots, \qquad |\tilde{\Delta}_F'(p)| \leqq C'\exp(|p^2|^\alpha), \quad \cdots, \quad \alpha < \tfrac{1}{2},$$

i.e., $\lim_{|p^2|\to\infty} |\rho(p^2)\exp(-\epsilon|p|) = 0$, *etc., for any* $\epsilon > 0$. *In particular, if in an unrenormalizable theory* $\rho(m^2)$ *has the form* $\rho(m^2) \approx \sum_0^\infty a_n m^{2n}$, *then causality implies that the coefficients* a_n *satisfy the relation*

$$(2.2)\qquad \limsup_{n\to\infty} \sqrt[n]{|a_n|(2n)!} = 0.$$

For the proof it suffices to note that if $\rho(p^2)$ would grow faster than allowed by (2.1), $\Delta_+'(x)$, etc., would develop space-like causality violating cuts in the complex x^2-plane. To see directly how in a causal unrenormalizable theory the condition (2.2) comes out, let us assume that $\rho(m^2)$ has the form $\rho(m^2) = 2\pi\sum_0^\infty 4^{-n}a_n(-m^2)^{n-1}$, $a_0 = 1/(-1)!$. Then, in the case of massless particles considered here for simplicity, we see from (1.1) that the commutator $\Delta'(x) = \Delta_+'(x) - \Delta_+'(-x)$ takes the form $\Delta'(x) = \sum_0^\infty a_n(n-1)!\,\epsilon(x_0)\delta^{(n)}(x^2)$, and for its Fourier transform one obtains

$(\epsilon(x_0) = \theta(x_0) - \theta(-x_0))$

$$\tilde{\Delta}'(p) = 2\pi \sum_0^\infty a_n (p^2)^{n-1} \theta(p^2) \epsilon(p_0).$$

Since $\epsilon(x_0)\delta^{(n)}(x^2)$ is the boundary value of an analytic function,

$$\epsilon(x_0)\delta^{(n)}(x^2) = \frac{(-i)^n n!}{2\pi i} [(x^2 - ix_0\, 0)^{-n-1} - (x^2 + ix_0\, 0)^{-n-1}],$$

it follows that the functional Δ' (symbolically, $\Delta'\langle\phi\rangle = \int_{-\infty}^{\infty} \Delta'(x)\phi(x)\, d^4x$) can be written as an analytic functional, viz.,

$$\Delta'\langle\phi\rangle = \frac{1}{2\pi i} \int_C dx_0 \int_{-\infty}^{\infty} G(x)\phi(x), \tag{2.3}$$

where

$$G(x) = \sum_0^\infty a_n n!\,(n-1)!\,(-x^2)^{-n-1}.$$

Here, C is a closed contour in the complex x_0-plane (Fig. 2) around the lightcone $x_0 = \pm|\mathbf{x}|$, which lies entirely within the region of convergence of the series G and which can be taken to be any circle enclosing the singular points ($x_0 = \pm|\mathbf{x}|$) of G. The support of Δ' is the smallest closed region to which the contour C can be shrunk. In case of massless particles, Δ' must have the lightcone as its support in virtue of the causality postulate. Hence, in order that the radius of C can be shrunk to the points $x_0 = \pm|\mathbf{x}|$ and be made arbitrarily small, the a_n must satisfy the condition (2.2) which leads immediately to the high energy bound (2.1). As has been indicated in §1, this means that the two-point functions have essential singularities on the lightcone.

3. Time-ordered and nontime-ordered functions. 1. The nontime-ordered (or Wightman) functions f_+ ($= \Delta_+'$, etc.) encountered are boundary values

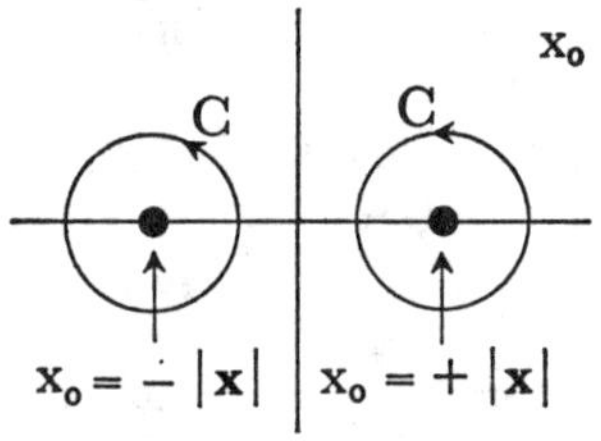

FIG. 2. *The contour in formula* (2.3)

of analytic functions of the type $(x^2 - ix_0 0)^{-\lambda}$, λ arbitrary, and series

$$f_+(x) = \sum_1^\infty a_n (x^2 - ix_0 0)^{-\eta_n}, \tag{3.1}$$

where $\eta_n \to \infty$ as $n \to \infty$, in the case of massless particles. Logarithmic terms may be included by differentiating with respect to some of the η_n. The associated generalized functions

$$\begin{aligned}(x^2 - ix_0 0)^{-\lambda}\langle\phi\rangle &= \lim_{\eta \to 0} ((x - i\eta)^2)^{-\lambda}\langle\phi\rangle \\ &= \lim_{\eta \to 0} \int_{-\infty}^{\infty} ((x - i\eta)^2)^{-\lambda} \phi(x - i\eta)\, d^4x\end{aligned}$$

are regular functionals (η time-like), (3.1) being assumed to converge in the spaces Z' or $S_\alpha'^\beta$, and the a_n satisfy the requirement (2.2). These Wightman functions cannot be rotated to Euclidean space.

2. Time-ordered functions f_F ($= \Delta_F'$, etc.) (propagators and Bethe-Salpeter amplitudes) are, in the massless case, built up by series of generalized functions of the type ($\eta_{\mu\nu}x^\mu x^\nu$ positive-definite)

$$(x^2 - i0)^{-\lambda} = \lim_{\eta_{\mu\nu} \to 0} (x^2 - i\eta_{\mu\nu}x^\mu x^\nu)^{-\lambda}, \qquad \lambda \text{ arbitrary},$$

e.g.,

$$f_F(x) = \sum_1^\infty a_n (x^2 - i0)^{-\eta_n}, \quad \eta_n \to \infty \text{ as } n \to \infty. \tag{3.2}$$

They can be viewed as boundary values of nonanalytic functions singular at the origin $x_0 = \mathbf{x} = 0$ of the lightcone and need particular definition: In analogy with what has been said in the preceding paper [1], the generalized function $(x^2 - i0)^{-\lambda}$ is defined by the functional (see [2], [3])

$$(x^2 - i0)^{-\lambda}\langle\phi\rangle = \operatorname*{Res}_{z=0} \left[\frac{1}{z} \int_{-\infty}^{\infty} (x^2 - i0)^{-\lambda} \left(\frac{-\mathbf{a}^2}{x^2 - i0}\right)^z \phi(x)\, d^4x\right], \tag{3.3}$$

where $\mathbf{a}$ is an arbitrary scaling parameter of dimension of length. The left-hand side of (3.3) depends on $\mathbf{a}$ if and only if $\lambda = n$, $n = 2, 3, \cdots$, i.e., if the classical, formal integral $\int_{-\infty}^{\infty} (x^2 - i0)^{-\lambda}\phi(x)\, d^4x$ exhibits logarithmic divergences. If $\lambda \neq n$, $n = 2, 3, \cdots$, then the right-hand side of (3.3) simply represents the functional obtained by continuing analytically in λ the regular functional $(x^2 - i0)^{-\lambda}\phi$ from $\operatorname{Re}\lambda < 0$ to the complex λ-plane. In (3.3) one first has to evaluate the x-integral for $\operatorname{Re} z$ sufficiently large negative, then to continue the resulting function of z to a region containing the origin of the z-plane and finally to apply the residue. If $\operatorname{Re}\lambda \geqq 1$, the Fourier transform of $(x^2 - i0)^{-\lambda}$ follows by replacing $\phi(x)$ in (3.2) by $\exp(ipx)$, in particular,

$$F[(x^2 - i0)^{-\lambda}] = -i(4\pi)^2(-4)^{-\lambda}\Gamma(2-\lambda)(-p^2 - i0)^{\lambda-2}/\Gamma(\lambda),$$

$$\lambda \neq 2, 3, \cdots,$$

$$F[(x^2 - i0)^{-n}] = \frac{i(4\pi)^2(-p^2 - i0)^{n-2}}{4^n(n-1)!(n-2)!} \tag{3.4}$$

$$\cdot\left[\log\left(\frac{\mathrm{a}^2(-p^2 - i0)}{4}\right) - \psi(n) - \psi(n-1)\right],$$

$$n = 2, 3, \cdots,$$

where $\psi(n) = \Gamma'(n)/\Gamma(n)$. Hence, two determinations of $(x^2 - i0)^{-n}$, $n = 2, 3, \cdots$, corresponding to two different scaling parameters a, a', differ from one another by a multiple of $\Box^{n-2}\,\delta(x)$:

$$(x^2 - i0)^{-n}|_{\mathrm{a}} - (x^2 - i0)^{-n}|_{\mathrm{a}'} = \text{const.} \log(\mathrm{a}^2/\mathrm{a}'^2)\,\Box^{n-2}\,\delta(x). \tag{3.5}$$

Consequently, $f_F(x)$ is determined, in case of integer η_n, only up to a term

$$f_0(x) = \log(\mathrm{a}^2/\mathrm{a}'^2)\sum_0^\infty c_n\,\Box^n\delta(x).$$

Since the a_n satisfy the causality condition (2.2), so do the c_n, and the series f_0 is localized at the origin of the lightcone. (As before, we assume f_F to be an element of Z' or $Z'\{M_q\}$.)

We wish to stress, that the classical homogeneous function $(x^2 - i0)^{-n}$ becomes nonhomogeneous if interpreted as a generalized function (as is immediately inferred, e.g., from (3.4)) if the parameter a is kept fixed. However, homogeneity is preserved if the length a is subject to the same dilatation $\mathrm{a} \to \alpha\mathrm{a}$ as the variable x, $x \to \alpha x$, i.e.,

$$((\alpha x)^2 - i0)^{-n}|_{\alpha\mathrm{a}} = \alpha^{-2n}(x^2 - i0)^{-n}|_{\mathrm{a}}.$$

a must be introduced entirely for physical reasons: Since x has dimension of length, p in (3.4) has the dimension of a reciprocal length, and if a were disregarded, the argument in the logarithm would have a dimension; this cannot be allowed [2], [3].

The formula (3.3) can be generalized as follows: Let $g(x^2 - i0)$ be a function having an algebraic singularity on the lightcone. Then the associated generalized function is represented by

$$g(x^2 - i0)\langle\phi\rangle = \operatorname*{Res}_{z=0}\left[\frac{1}{z}\int_{-\infty}^{\infty} g(x^2 - i0)\left(\frac{-\mathrm{a}^2}{x^2 - i0}\right)^z \phi(x)\,d^4x\right], \tag{3.6}$$

which upon Fourier transformation yields the following generalized Bochner formula [3],

$$F[g(x^2 - i0)] = -4i\pi^2 \cdot \operatorname*{Res}_{z=0}\left[\frac{1}{z(-p^2 - i0)^{1/2}} \int_0^\infty \tau^2 \left(\frac{-a^2}{\tau^2}\right)^z g(-\tau^2) \mathcal{J}_1(\tau(-p^2 - i0)^{1/2})\, d\tau\right]. \tag{3.7}$$

Products of nontime-ordered generalized functions are defined by convolution of their Fourier transforms, for these Fourier transforms form a convolution algebra. Products of time-ordered functions $g(x^2 - i0)$, $h(x^2 - i0)$ are defined in terms of the formula (3.6) according to

$$g(x^2 - i0)\cdot h(x^2 - i0)\langle\phi\rangle = \operatorname*{Res}_{z=0}\left[\frac{1}{z}\int_{-\infty}^{\infty} g(x^2 - i0)h(x^2 - i0)\left(\frac{-a^2}{x^2 - i0}\right)^z \phi(x)\, d^4x\right]. \tag{3.8}$$

Thus, for example, the propagator product $(x^2 - i0)^{-\alpha}\cdot(x^2 - i0)^{-\beta}$ equals the generalized function $(x^2 - i0)^{-\alpha-\beta}$. Further details on product theories may be found in [3]. Convolutions of singular, time-ordered functions may be formed in an analogous way. In particular, we obtain for any λ ($*$ denotes the convolution product)

$$(x^2 - i0)^{-1} * (x^2 - i0)^{-\lambda} = i\pi^2 \operatorname*{Res}_{z=0}\left[\frac{(x^2 - i0)^{-\lambda+1-z}a^{2z}}{z(1 - \lambda - z)(2 - \lambda - z)}\right]. \tag{3.9}$$

This can be proved by Fourier transformation and application of (3.4).

The series of the type (3.1) and (3.2) are defined as functionals by term-by-term application of the above rules. While nontime-ordered series are identical with the corresponding closed form expressions, time-ordered series represent generalized functions per se. These, in general, would not correspond to linear continuous functionals if viewed as closed form expressions (e.g., exp $(\sqrt{g}/(x^2 - i0))$ is defined by its power series expansion).

4. Generalized Lehmann representation. Having defined time-ordered functions we are now in a position to give a meaning to the spectral representation (1.1) for $\Delta_F'(x)$ (that of $\Delta_+'(x)$ does not present any problem!). Indeed, $\Delta_F'(x)$ is given meaning by means of the formula

$$\Delta_F'\langle\phi\rangle = \operatorname*{Res}_{z=0}\left[\frac{1}{z}\int_{-\infty}^{\infty}\left\{\int_0^\infty \rho(m^2)\Delta_F(x, m^2)\right\}\left(\frac{-a^2}{x^2 - i0}\right)^z \phi(x)\, d^4x\right], \tag{4.1}$$

from which it follows that

$$\Delta_F'\langle\phi\rangle = \operatorname*{Res}_{z=0}\left[\frac{1}{z}\int_0^\infty \rho(m^2)(a^2m^2)^z\{\Delta_F(x, m^2)\phi\}\, dm^2\right] \tag{4.2}$$

for $\rho(m^2) = (m^2)^\lambda \log^\mu m^2$, $\lambda \geqq 0$, $\mu = 0, 1, 2, \cdots$. Taking the Fourier transform of (4.2) we arrive at the unsubtracted spectral representation (generalized Lehmann representation)

$$\tilde{\Delta}_F'(p) = 2i \operatorname*{Res}_{z=0} \left[\frac{1}{z} \int_0^\infty \frac{\rho(m^2)(\mathrm{a}^2 m^2)^z}{p^2 - m^2 + i0} \right], \tag{4.3}$$

and a similar representation holds for $\tilde{S}_F'(p)$, the nontime-ordered representations remaining unaltered. In (4.3) one has, in the case of increasing tempered $\rho(m^2)$, first to perform the m^2-integral for Re z sufficiently large negative, then to continue the resulting function in z and finally to apply the residue. In the case of infinite series $\rho \approx \sum_0^\infty a_n (m^2)^{\eta_n} \log^\mu m^2$, the spectral representation is constructed by term-by-term application of the residue procedure.

For the absorptive part of $\tilde{\Delta}_F'(p)$ we obtain, of course,

$$2i \operatorname*{\mathcal{I}m}_{p^2 \geqq 0} \tilde{\Delta}_F'(p) \equiv \operatorname*{disc}_{p^2 \geqq 0} \tilde{\Delta}_F'(p) = 4\pi\rho(p^2),$$

which shows that these absorptive parts across physical cuts are independent of the arbitrary scaling parameter **a**. Similarly one can show that on-mass-shell S-matrix elements are independent of **a**.

5. Integral equations for unrenormalizable interactions. The integral equation for time-ordered two-point functions or Bethe-Salpeter amplitudes f_F take in the local (string or ladder) approximation the form

$$f_F(x) = f_F^0(x) + g \int_{-\infty}^{\infty} G_F(x - x')V_F(x')f_F(x')\, d^4x', \tag{5.1}$$

where G_F is a Green's function, $G_F(x) \approx (x^2 - i0)^{-1}$, $x^2 \to 0$, in the case of massless particles, and V_F is an interaction kernel which is singular,

$$V_F(x) \approx (x^2 - i0)^{-\alpha}, \qquad x^2 \to 0, \quad \alpha > 1, \tag{5.2}$$

in the case of unrenormalizable interactions. f_F^0 is a free field function and g a coupling constant. In virtue of the singular kernel (and because of the generalized function product $V_F \cdot f_F$ involved), (5.2) is neither of Fredholm type nor does it have a classical (or even distribution) analogue. Without giving a proper interpretation to (5.1) one only finds by formal iteration a "solution" of the type $f_F(x) = f_F^0(x) + g\infty + g^2\infty^2 + \cdots$, which merely represents the infinite set of divergent Feynman diagrams (whose order of divergence increases with that of g^n) and which have been summed into the equation. (Figure 3 indicates the diagrams belonging to (5.1) in case of a scattering process within a four-fermion theory.) As a concrete model to (5.1) we refer to the Bethe-Salpeter equation [2]

$$\psi_F(x) = 1 - (ig/\pi^2) \int_{-\infty}^{\infty} \frac{1}{(x - x')^2 - i0} \frac{1}{(x'^2 - i0)^3} \psi_F(x')\, d^4x \tag{5.3}$$

and to the propagator equation

$$S_F'(x) = S_F(x) - (g/4)\int_{-\infty}^{\infty} S_F(x-\mu) \tag{5.4}$$
$$\cdot \frac{1}{((u-v)^2 - i0)^3} S_F'(u-v) S_F(v)\, d^4\mu\, d^4v$$

in a 4-fermion theory (with $S_F(x) = -(i\gamma\partial + m)\Delta_F(x)$).

Let us now interpret (5.1) as a convolution equation in the sense of generalized functions. Then the equation can obviously be solved by straightforward iteration using the concepts introduced in §3. That is to say, (5.1) has a finite perturbative solution in Minkowski space and, at the same time, the generalized function interpretation also renders the Feynman diagrams associated with the different orders altogether finite. For example, (5.3) can immediately be iterated by the help of (3.9) with the result

$$\psi_F(x) = \sum_0^{\infty} g^n(x^2 - i0)^{-2n}/(2n)! \equiv \cosh\left(\sqrt{g}/(x^2 - i0)\right), \tag{5.5}$$

and similarly, the perturbative solution of (5.4) is found to be

$$S_F'(x) = S_F(x)(-\sqrt{g}/2(x^2 - i0))^{-1} I_1(-\sqrt{g}/(x^2 - i0)). \tag{5.6}$$

These series converge in Z', $Z'\{M_q\}$ or $S_\alpha'^\beta$, but in analyzing the general equation (5.1) one has to assume the series solution f_F convergent since at present no general theory for equations of this type is available. According to what had been said in §3, the series solutions in general are determined only up to a generalized function $f_0 = \sum_0^\infty a_n \Box^n \delta(x)$ localized at the origin of the lightcone, unless the scaling parameter a is fixed. That the solutions so obtained are the correct ones may already be inferred from the differential equations associated with (5.3) and (5.4), viz., $\Box\psi_F(x) = -4g(x^2 - i0)^{-3}\psi_F(x)$ and $(i\gamma\vec{\partial} - m)S_F'(i\gamma\overleftarrow{\partial} - m) = -g(x^2 - i0)^{-3}S_F'$, respectively. These are differential equations of a rather new type, containing singular generalized functions as coefficients. While in the second case the solution is uniquely determined by the boundary condition $S_F' \to S_F$ as $x_0 \to \infty$ and, in fact, given by (5.6), no boundary condition is known for

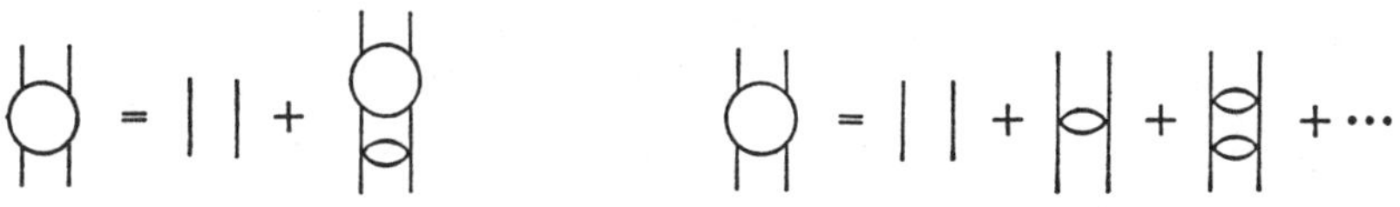

$$f_F = f_F^0 + gG_F * (V_F f_F)$$

FIG. 3. *The Feynman diagram representing* (5.1) *and its graphical solution* f_F *(Bethe-Salpeter case)*

the first (Bethe-Salpeter) case so that the general invariant solution of that differential equation is given by $\psi_F = \exp(-\sqrt{g}/(x^2 - i0)) + A \exp(+\sqrt{g}/(x^2 - i0))$, A being an arbitrary constant.

Things are much simpler in the case of the integral equation

$$f_+(x) = f_+^0(x) + g \int_{-\infty}^{\infty} G_R(x - x')V_+(x')f_+(x')\, d^4x' \tag{5.7}$$

for the corresponding nontime-ordered function $f_+(x)$, where f_+^0 is the free function, $G_R(x) \approx \theta(x_0)\delta(x^2)$ in the case of massless particles and $V_+(x) \approx \text{const.}\ (x^2 - ix_0 0)^{-\alpha}$, $\alpha > 1$. Indeed, since the quantities involved are boundary values of analytic functions, (5.7) is of Fredholm type and possesses a perturbative solution in the classical sense. The nontime-ordered equations associated with (5.3) and (5.4) are, respectively,

$$\begin{aligned}\psi_+(x) &= 1 \\ &\quad - (2g/\pi) \int_{-\infty}^{\infty} \theta(x_0 - x_0')\delta((x - x')^2)\, \frac{1}{(x'^2 - ix_0 0)}\, \psi_+(x')\, d^4x',\end{aligned} \tag{5.8}$$

$$\begin{aligned}S_+'(x) &= S_+(x) \\ &\quad - g \int_{-\infty}^{\infty} S_R(x - u)\, \frac{1}{((u - v)^2 - ix_0 0)^3}\, S_+'(u - v)S_A(v)\, d^4u\, d^4v,\end{aligned} \tag{5.9}$$

where $S_R(x) = -i\gamma\partial\theta(+x_0)\delta(x^2)/2\pi$, $S_A(x) = -i\gamma\partial\theta(-x_0)\delta(x^2)/2\pi$. The iterative solutions are

$$\psi_+(x) = \cosh(\sqrt{g}/(x^2 - ix_0 0)) \tag{5.10}$$

and

$$S_+'(x) = S_+(x)(-\sqrt{g}/2(x^2 - ix_0 0))^{-1} I_1(-\sqrt{g}/(x^2 - ix_0 0)), \tag{5.11}$$

respectively. They follow readily by utilizing the formula

$$(x^2 - ix_0 0)^{-1} * [-(x^2 - ix_0 0)]^{-\lambda} = \frac{-i\pi^2[-(x^2 - ix_0 0)]^{-\lambda-1}}{(\lambda - 1)(\lambda - 2)},$$

which is proved in a way similar to (3.9).

It is immediately seen that the Minkowski space iterative solutions of the time-ordered equations are connected with the corresponding nontime-ordered solutions by the relation

$$f_F(x) = f_+(\pm x), \qquad x_0 \gtrless 0. \tag{5.12}$$

And for the absorptive parts, i.e., for the discontinuities across the physical cuts in momentum space, we obtain

$$2i \underset{p^2 \geqq 0}{\mathcal{I}m}\, f_F(p) \equiv \underset{p^2 \geqq 0}{\text{disc}} = 2if_+(p) \tag{5.13}$$

in agreement with the Cutkosky rule.

To summarize: by properly defining the singular quantities of field theory in terms of Lorentz-invariant generalized functions we have been able to give a meaning to the equations of unrenormalizable interactions and have obtained finite solutions rather than "an infinite set of divergent integrals whose degree of divergence increases with the order of the diagrams". Furthermore, one can show that the product formalism sketched in §3, applied to the individual Feynman diagrams, yields results identical with the solutions of the integral equations. While the time-ordered solutions are determined only up to a term $f_0 = \sum_0^\infty c_n \Box^n \delta(x)$ localized at the origin of the lightcone (according to the causality postulate) due to the arbitrary scaling parameter **a**, this term does not influence physical observables within the approximations to the interaction considered so far.

6. Peratization. Time-ordered equations can also be defined by means of the so-called complex extension method [2], [3] of which Euclidean rotations in x_0- and $\mathbf{x}$-planes are special cases. The point is that (e.g., by rotating the real integration contour $-\infty < x_0 < \infty$ to the imaginary axis) by complex extension one associates with the Minkowski space equation (5.1) a Euclidean equation ($x_e'^2 = x_{e0}'^2 + \mathbf{x}_e'^2$, $x_e^2 = x_{e0}^2 + \mathbf{x}_e^2$),

$$(6.1)\quad f_F^e(x_e) = f_F^{0e}(x_e) + g \int_{-\infty}^{\infty} G_F^e(x_e - x_e') V_F^e(x_e') f_F^e(x_e')\, d^4x_e',$$

so that, for example, in the Bethe-Salpeter case (5.3), the associated complex extended, Euclidean equation reads

$$(6.2)\qquad \psi_F^e(x_e) = 1 - \frac{ig}{\pi^2} \int_{-\infty}^{\infty} \frac{1}{(x_e - x_e')^2} \frac{1}{(x_e'^2)^3} \psi_F^e(x_e')\, d^4x_e'.$$

The Euclidean equation, however, admits two interpretations:

(i) Interpreted as an equation in the sense of generalized functions (i.e., $(x_e'^2)^{-\lambda}\langle\phi\rangle$ = generalized function = $\mathrm{Pf}\,(x_e'^2)^{-\lambda}\langle\phi\rangle$ (pseudofunction)), the equation has a perturbative solution which, after passing to Minkowski space, coincides with the original Minkowski space solution. For instance, (6.2) can be iterated by means of the formula

$$(r^2)^{-1} * (-r^2)^{-\lambda} = -i\pi^2 \operatorname*{Res}_{z=0} \left[\frac{(-r^2)^{-\lambda+1-z} \mathrm{a}^{2z}}{z(\lambda - 1 + z)(\lambda - 2 + z)} \right],$$

$$r^2 = x_{e0}^2 + \mathbf{x}_e^2,$$

to yield $\psi_F^e(x_e) = \cosh(\sqrt{g}/x_e'^2) \to \cosh(\sqrt{g}/(x^2 - i0))$ in agreement with (5.5).

(ii) Interpreted as a classical, though non-Fredholm equation and requiring the solution to be integrable in the ordinary sense, the equation considered has either no solution at all (which happens in the propagator case) or it has one integrable solution (which happens in the Bethe-Salpeter case). The latter solution, in general, is nonanalytic in the coupling constant g. For example, (6.2), interpreted in this sense, has the solution $\psi_F^e(x_e)_{\text{classical}} = \exp(-\sqrt{g}/x_e^2)$. Passing to Minkowski space, one can show that out of this classical solution two Minkowski space solutions emerge, i.e., $\psi_F^e(x) = \exp(-\sqrt{g}/(x^2 - i0)) + A \exp(\sqrt{g}/(x^2 - i0))$ with an arbitrary constant A, since this process is not uniform.

Peratization [6] is a technique for obtaining such nonanalytic solutions in an iterative way. However, as we have seen, this must be based upon the interpretation (ii) of the Euclidean equation. This Euclidean interpretation of the original equation in Minkowski space distorts the original interaction concept. The solution, nonanalytic in g, clearly does not preserve the necessarily perturbative absorptive parts: the relation (5.12) and (5.13) is not satisfied with $f_{F,+} = \psi^e_{f,+,\text{classical}}$. Although such nonperturbative solutions possess some very interesting physical aspects (e.g., yielding a damping of scattering amplitudes at high energies), it is not yet clear whether to reject them or not in the Bethe-Salpeter case.

7. Conclusions. We have shown that the application of the techniques of Lorentz-invariant generalized functions permits one to give a meaning to thus far undefined quantities in unrenormalizable quantum field theories and to obtain solutions to a class of unrenormalizable equations.

Although our approach is rather incomplete and raises more problems than have been solved, the results indicate strongly that the techniques of Lorentz-invariant generalized functions may provide the clue in a systematic study of the difficult problems of unrenormalizable field theory.

Acknowledgments. One of the authors, W. Güttinger, wishes to express his sincere gratitude to the organizers of the SIAM 1966 Fall Meeting for the kind invitation. He is much indebted to many colleagues participating at the conference for valuable discussions, in particular, to A. S. Wightman, A. Jaffe, A. H. Zemanian, R. Haag, R. Newcomb and H. Bremermann. Thanks are also due to the Deutsche Forschungsgemeinschaft for a research grant.

REFERENCES

[1] W. GÜTTINGER, *Generalized functions in elementary particle physics and passive system theory: Recent trends and problems*, this Journal, 15 (1967), pp. 964–1000.

[2] W. GÜTTINGER AND E. PFAFFELHUBER, *Dynamics of unrenormalizable interactions in Minkowski and Euclidean space*, Nuovo Cimento, to appear.

[3] W. GÜTTINGER, *Generalized functions and dispersion relations in physics*, Fortschr. Physik, 14 (1966), pp. 483–602.

[4] S. SCHWEBER, *An Introduction to Relativistic Quantum Field Theory*, Harper and Row, New York, 1962.

[5] A. S. WIGHTMAN, *Quantum field theory in terms of vacuum expectation values*, Phys. Rev., 101 (1956), pp. 860–868.

[6] W. GÜTTINGER, R. PENZL AND E. PFAFFELHUBER, *Peratization of unrenormalizable interactions*, Ann. Physics, 33 (1965), pp. 246–271.

[7] B. SCHROER, *Some remarks on non-renormalizable quantum field theories*, High Energy Physics and Elementary Particles, A. Salam, ed., International Atomic Energy Agency, Vienna, 1965, pp. 295–603.

A GENERALIZATION OF THE PALEY-WIENER-SCHWARTZ THEOREM*

ARTHUR M. JAFFE†

We are interested in the question of how to characterize the Fourier transform of generalized functions [1] which can be strictly localized in a compact region of R^n. We start by defining a set of translation invariant, strictly local test function spaces $\mathfrak{L}(R^n)$. These spaces are subsets of $\mathfrak{D}(R^n)$, the space of all infinitely differentiable functions with compact support. Each space is characterized by an entire analytic function $g(t)$:

$$g(t) = \sum_{r=0}^{\infty} c_{2r} t^r, \qquad c_{2r} \geqq 0, \quad c_0 \neq 0. \tag{1}$$

For any bounded, open region $\mathfrak{O}$ in R^n, define the local functions $\mathfrak{L}(\mathfrak{O})$ as the countably normed set of functions $f(x)$ with support in $\mathfrak{O}$, for which all the norms

$$\| f \|_s = \sum_{r=0}^{\infty} c_{2r} \sup_{x \in \mathfrak{O}} | \Delta^{r+s} f(x) |$$

are finite. Here $\Delta = \sum_{j=1}^{n} \partial^2/\partial x_j^2$ is the n-dimensional Laplace operator. Let $\mathfrak{L}(R^n)$ be the inductive limit of spaces $\mathfrak{L}(B_R)$ corresponding to the n-balls B_R of radius R, centered at the origin.

THEOREM 1. *Either $\mathfrak{L}(R^n)$ contains only the function $f \equiv 0$, or else $\mathfrak{L}(R^n)$ is dense in $\mathfrak{D}(R^n)$. A necessary and sufficient condition that $\mathfrak{L}(R^n)$ be non-trivial is that either*

$$\int_0^{\infty} \frac{\log g(t^2)}{1 + t^2} \, dt < \infty, \tag{2a}$$

or

$$\sum_{r=0}^{\infty} \sup_{s \geqq 0} (c_{2r+2s})^{1/(2r+2s)} < \infty. \tag{2b}$$

DEFINITION. A generalized function T is *strictly localizable* in $\mathfrak{O}$ if T is a continuous linear functional on some $\mathfrak{L}(R^n)$ and T has its support in $\mathfrak{O}$.

* Received by the editors November 17, 1966. Contributed at the Symposium on "The Applications of Generalized Functions" sponsored by the Air Force Office of Scientific Research at the 1966 Fall Meeting of Society for Industrial and Applied Mathematics held at the State University of New York at Stony Brook, September 12–14, 1966.

† Department of Mathematics, Stanford University, Stanford, California. This work was supported by the Air Force Office of Scientific Research while the author was a National Academy of Sciences-National Research Council fellow.

In other words, T is an element of some $\mathfrak{L}'(R^n)$ and $T(f) = 0$ for every $f \in \mathfrak{L}(R^n)$ which has support in $R^n - \mathfrak{D}$.

THEOREM 2. *A generalized function T is strictly localizable in B_R if and only if it is the Fourier transform of an entire function F for which*

$$|F(p + iq)| \leqq g(\|p\|^2)e^{R\|q\|}, \tag{3}$$

where

$$p \in R^n, \qquad q \in R^n, \qquad \|p\|^2 = \sum_{j=1}^{n} p_j^2,$$

and where $g(t)$ is an entire function satisfying (1) *and* (2).

Remark. If f in (3) is a polynomial, T is a Schwartz distribution and the theorem reduces to the usual Paley-Wiener-Schwartz result [3].

Proof of Theorem 1. It follows from the theory of quasi-analytic classes of functions [2] that if $\mathfrak{L}(R^1)$ is nontrivial then it contains a nontrivial, positive function. Convolution by $\mathfrak{D}(R^1)$ leaves $\mathfrak{L}(R^1)$ invariant, from which it easily follows that $\mathfrak{L}(R^1)$ is dense in $\mathfrak{D}(R^1)$. By taking tensor products, $\mathfrak{L}(R^n)$ is dense in $\mathfrak{D}(R^n)$. Hence if $\mathfrak{L}(R^1)$ is nontrivial, $\mathfrak{L}(R^n)$ is dense in $\mathfrak{D}(R^n)$. Conversely, if f is a nontrivial element of $\mathfrak{L}(R^n)$, fixing $(n - 1)$ variables yields a nontrivial element of $\mathfrak{L}(R^1)$. This proves the first statement of the theorem. The space $\mathfrak{L}(R^1)$ is nontrivial if and only if it is contained in no quasi-analytic class. From the Denjoy-Carleman-Ostrowski theorem [2] we infer that this is equivalent to (2a) or (2b).

Proof of Theorem 2. Let $f_\epsilon(x) = \epsilon^{-n}f(x/\epsilon)$ be a positive element of $\mathfrak{L}(R^n)$ whose integral is one. Then for any $T \in \mathfrak{L}'(R^n)$ the regularization $T_\epsilon = T * f_\epsilon$ is an infinitely differentiable function which converges in $\mathfrak{L}'$ to T as $\epsilon \to 0$. From the Paley-Wiener theorem for functions, T_ϵ has its support in $B_{R+\epsilon M}$, which in the limit $\epsilon \to 0$ gives the desired result for T. Conversely, if T is strictly localized in B_R, an estimate of $T(e^{ipx})$ leads to (3).

REFERENCES

[1] I. M. GELFAND AND G. E. SILOV, *Generalized Functions*, vols. 1, 2, Academic Press, New York, 1964.

[2] S. MANDELBROJT, *Séries adhérentes, régularization des suites, applications*, Gauthier-Villars, Paris, 1952.

[3] L. SCHWARTZ, *Théorie des distributions*, Hermann, Paris, 1950.

PHYSICALLY MOTIVATED DEFINITIONS OF DISTRIBUTIONS*

T. P. G. LIVERMAN†

1. Introduction. Given a vector space $\mathfrak{B}$ of C^∞ functions $\varphi: R^n \to C$, equipped with a topology relative to which differentiation is a continuous operation and the space is complete, the set $\mathfrak{B}'$ of distributions is the vector space of continuous linear functionals $F: \mathfrak{B} \to C$. This, roughly, is the original approach of L. Schwartz to generalized functions [17] extended in [3], [4] and also exposed in [2], [6], [13], [19].

A less abstract definition of distributions, suggested by J. Mikusinski and first carried out by G. Temple, is patterned after Cantor's definition of real numbers as equivalence classes of sequences of rationals. Starting with an appropriate vector space $V(\mathfrak{B})$ of functions $f: R^n \to C$ such that the integrals $\int_{R^n} f(x)\varphi(x)\,dx$ exist, $\forall\varphi \in \mathfrak{B}$, one calls *fundamental* (on $\mathfrak{B}$) any sequence $\{f_\nu\}$ of such functions for which $\lim_{\nu\to\infty} \int f_\nu(x)\varphi(x)\,dx$ exists at every $\varphi \in \mathfrak{B}$. This limit is the value $\langle F, \varphi\rangle$ of a functional $F: \mathfrak{B} \to C$. The set of these F turns out to be precisely the class $\mathfrak{B}'$ of distributions. This procedure for various $\mathfrak{B}$ is found in [9], [10], [14], [18]; that it is equivalent to the continuous linear functional approach follows from Schwartz's theory [17, vol. I, p. 75; vol. II, p. 95]. It has also been proved by elementary (i.e., classical analysis) methods [10], [15]. $\mathfrak{B}$ is often called the *test space* and its φ, *testing functions* of the class $\mathfrak{B}'$ of distributions.

In this paper it is shown that there is no need to test sequences $\{f_\nu\}$ at *every* $\varphi \in \mathfrak{B}$ to find out whether they are fundamental and define an $F \in \mathfrak{B}'$. Instead it suffices to confine oneself to those $\varphi \in \mathfrak{B}$ which are *probability densities*, i.e., which have $\varphi(x) \geqq 0$ and $\int_{R^n} \varphi(x)\,dx = 1$. This is done for the g.f. (generalized function) classes $\mathfrak{D}'$, $\mathcal{S}'$ considered in [2], [3], [17] as well as for the classes $S_{\alpha,A}$ defined the abstract way in [2], [3], [4].

In outline what is done here is this: given $\Delta(\mathfrak{B})$, the set of probability densities φ belonging to $\mathfrak{B}$ (where $\mathfrak{B}$ is any one of the test spaces $\mathfrak{D}(R^n)$, $\mathcal{S}(R^n)$, $S_{\alpha,A}(R^n)$ whose definition is recalled in §3), and given certain spaces $V(\mathfrak{B})$ of continuous functions described in §4, we call "basic" any sequence $\{f_\nu\}$ of $V(\mathfrak{B})$ functions such that the quantity $\langle F, \varphi\rangle = \lim \int f_\nu\varphi\,dx$ exists

* Received by the editors December 6, 1966, and in revised form January 19, 1967. Contributed at the Symposium on "The Applications of Generalized Functions" sponsored by the Air Force Office of Scientific Research at the 1966 Fall Meeting of Society for Industrial and Applied Mathematics held at the State University of New York at Stony Brook, September 12–14, 1966.

† George Washington University, Washington, D.C.

for every $\varphi \in \Delta(\mathfrak{B})$. The functional $F:\Delta(\mathfrak{B}) \to C$ thus defined is then (§§6, 7) shown to be a continuous linear functional on all of $\mathfrak{B}$ as well—in fact $\mathfrak{B}'$ is found to be the set of these F. From this we proceed to study the properties of sequences $\{F_\nu\}$ of $\mathfrak{B}'$ distributions, proving (§8) that the completeness and relative compactness properties of $\mathfrak{B}'$ are entirely determined by the values of the F_ν on the subset $\Delta(\mathfrak{B})$ of $\mathfrak{B}$.

The idea behind this treatment of distributions as functionals on probability densities is physical. It stems from an acceptance of the fact that, when one measures a physical magnitude $m(x)$ in the laboratory, random errors intrude upon the determination of the independent variable no less than they do on that of the dependent quantity m. Some remarks on this score precede the mathematical content of this report which begins with §3. Apropos the latter, let it be mentioned that the exposition is so couched as to require, by way of topology, acquaintance with no more than some rudiments. It is perhaps worth indicating, too, that the method of proof given here yields basic structural results (Corollaries 1, 2, 3) on distributions, without recourse to the Hahn-Banach theorem.

2. Physical background. The physical magnitude functions about which one theorizes are constructs—inventions of the mind designed to correlate (interpolate and predict) experimental facts in a consistent, systematic and, according to many, intellectually economical way. One much indulged in economy of thought consists in assuming these magnitudes to be the values of appropriately smooth functions $f(x, t)$ of space and/or time variables, which are appropriately smooth in the sense that, outside certain isolated points and boundary regions, they are continuous and have continuous derivatives up to such order p as one finds convenient to characterize their behavior in the small by means of ordinary or partial differential equations.

Traditionally, in classical physics at least, one supposes that a measurement of $f(x, t)$ actually yields a quantity $f(x, t) + e(x, t)$, where e is a random error with a probability density $\delta(e)$ such that the mean $\bar{e}(x, k) = \int_R e\delta(e)\, de = 0$ and the variance $\int_R e^2\delta(e)\, de$ is finite. The expected value of a measurement at $(x, t) \in R^3 \times R$ will in consequence be $f(x, t)$. This, it is generally agreed, is an oversimplification; after all, if a thermometer reading of a temperature f or a galvonometer reading of a current f are tainted with an error e, is it likely that the yardstick and chronometer readings which purport to give the location (x, t) to which $f + e$ applies are flawless?

To take such strictures into account we may suppose that a measurement intended at the location (x, t) actually occurs in $(\xi, \xi + d\xi) \times (\tau, \tau + d\tau)$ with probability $\varphi(\xi, \tau)\, d\xi\, d\tau$ (where $\int_{R^4} \xi\varphi(\xi, \tau)\, d\xi\, d\tau = x$ and

$\int_{R^4} \tau\varphi(\xi, \tau)\, d\xi\, d\tau = t$ give the mean location of (ξ, τ) as (x, t)) and that the expected value of a measurement of f intended at (x, t) is the weighted average $\langle f, \varphi \rangle = \int_{R^4} f(\xi, \tau)\varphi(\xi, \tau)\, d\xi\, d\tau$. (Heuristically, if there exist dependent variable error functions $e_k(\xi, \tau)$, and a particular one, e_n, were in effect during the experiment, the expected result would be $\langle f + e_n, \varphi \rangle = \langle f, \varphi \rangle + \langle e_n, \varphi \rangle$; but the functions e_k are random, and we assume the expected value of $\langle e_k, \varphi \rangle$ over the various k to be zero, in natural extension of the assumption $\bar{e}(x, t) = 0$.) $\langle f, \varphi \rangle$ is often called a "smeared" value of f, and φ a "smearing function".

So long as we take the attitude that all we can physically "see" of an f is the set of smeared values $\langle f, \varphi \rangle$, we must adopt mathematical rules for the manipulation of f that are consistent with observational facts. In particular is it necessary that we consider a new type of limit process for sequences $\{f_\nu\}$ of physical magnitude functions. Indeed, to say that $\lim_\nu f_\nu(x, t)$ exists at some point (x, t), or uniformly in some region of R^4, becomes a mathematical assertion which is not physically verifiable. Considerably more relevant operationally (in the sense of Bridgman [1]) is the statement: $\lim_\nu \langle f_\nu, \varphi \rangle$ exists for every φ. In fact, consistency demands that to the usual catalogue of constructs—the functions f of (x, t)—we add all functionals F on the set Σ of the φ whose values $\langle F, \varphi \rangle$ (this notation replaces $F(\varphi)$) are of the form $\langle F, \varphi \rangle = \lim_\nu \langle f_\nu, \varphi \rangle$. Since each f has $\langle f, \varphi \rangle = \lim_\nu \langle f_\nu, \varphi \rangle$ with $f_\nu = f$, $\nu = 1, 2, \cdots$, the f form a subset of Σ', the collection of all F.

The collection Σ' thus arrived at depends on the class Σ of smearing functions φ one starts out with and on the f_ν for which we define $\langle f_\nu, \varphi \rangle$. The choices made here are these: for Σ we take (see §4) various sets of $C^\infty(R^n)$ probability densities which are distinguished by their rate of decrease as $|x| \to \infty$, and for the f_ν we take sets $V(\Sigma)$ of $C(R^n)$ functions that yield $\int_{R^n} f_\nu(x)\varphi(x)\, dx < \infty$. These assumptions, which are physically reasonable (see §4, Remark), in each case lead to the conclusion that the corresponding Σ' is a set of generalized functions—in other words, the F defined as functionals on probability densities simply turn out to be continuous linear functionals on much larger sets $\mathfrak{B}$ of $C^\infty(R^n)$ functions that contain Σ as a proper subset.

Thus, by taking into account the presence of errors in the experimental determination of both the independent and dependent variables in physical phenomena, one is led to the conclusion that generalized, rather than ordinary, functions are the natural entities for the pencil and paper depiction of these phenomena. One consequence of this is that there is no harm in assuming whatever smoothness conditions one may wish, when deriving differential equation models; for consistency demands that these distribution solutions be granted equal relevance to the description of physical events. From the fact that only positive smearing functions with $\int \varphi = 1$ are required for the definition of the g.f.'s, it follows that the experimental confirmation or refutation of a differential equation model (e.g., $Ly(x, t) = f(x, t)$, where L is a linear differential operator) depends only on checks of the $\langle y, \varphi \rangle$ and $\langle f, \varphi \rangle$ of the output y and input f with such particular φ—not with the much vaster set $\mathfrak{V}$ of φ's demanded by the usual physical interpretation of distributions.

3. Test spaces and distributions. $[-l, l]$ stands for the closed interval $[-l, l] \times \cdots \times [-l, l] \subset R^n$, where l is a positive integer.[1] For $\partial^m/\partial x_i^m$ we write ∂_i^m and, given any n-tuple $\sigma = (\sigma_1, \cdots, \sigma_n)$ of nonnegative integers, we use the abbreviations ∂^σ for $\partial_1^{\sigma_1}\partial_2^{\sigma_2} \cdots \partial_n^{\sigma_n}$ and define $|\sigma|$ by $|\sigma| = \sigma_1 + \sigma_2 + \cdots + \sigma_n$. The carrier, or support, of a function $f: R^n \to C$, denoted by supp (f), is the closure of the set $\{x : f(x) \neq 0\}$, and f is called $C^\infty(R^n)$ if its domain is R^n, its values real or complex and $\partial^\sigma f$ is continuous on R^n for every σ with $|\sigma| \geqq 0$ ($\partial^0 f = f$ by convention). If $k = (k_1, \cdots, k_n)$ we put x^k for $x_1^{k_1} x_2^{k_2} \cdots x_n^{k_n}$.

$\mathfrak{D}[-l, l]$ represents the set of all $C^\infty(R^n)$ functions φ with values $\varphi(x) = \varphi(x_1, \cdots, x_n)$ having supp $(\varphi) \subset [-l, l]$. This is a vector space over the field C of complex numbers such that for every integer $p \geqq 0$ and every $\varphi \in \mathfrak{D}[-l, l]$ the quantity

$$\| \varphi \|_p = \sup_{|\sigma| \leqq p} \sup_{x \in R^n} | \partial^\sigma \varphi(x) | \tag{3.1}$$

is finite [2], [3], [4], [17]. Here the first supremum symbol on the right-hand side is understood to be taken over all σ for which $|\sigma| \leqq p$, the second over all $x \in R^n$.

$\mathcal{S}(R^n)$ denotes the vector space of all $C^\infty(R^n)$ functions φ of "rapid decrease" [2], [4], [17], [19]. Specifically, $\varphi \in \mathcal{S}(R^n)$ if and only if $\varphi \in C^\infty(R^n)$ and to every pair of n-tuples $\sigma = (\sigma_1, \cdots, \sigma_n)$ and $k = (k_1, \cdots, k_n)$ of nonnegative integers there corresponds a constant $C_{k\sigma}$ such that

$$\sup_x | x^k \partial^\sigma \varphi(x) | \leqq C_{k\sigma} . \tag{3.2}$$

Setting

$$M_p(x) = (1 + |x_1|)^p \cdots (1 + |x_n|)^p \tag{3.3}$$

for $p = 0, 1, \cdots$, it follows from (3.2) (see [4, p. 81]) that for every

[1] It is always clear from the context whether l stands for an integer or for the n-tuple $(l, l, \cdots, l)$.

$\varphi \in \mathcal{S}(R^n)$ the quantities $\| \varphi \|_p$ defined by

$$\| \varphi \|_p = \sup_{|\sigma| \leqq p} \sup_{x \in R^n} M_p(x) | \partial^\sigma \varphi(x) | \tag{3.4}$$

are finite.

Given $\alpha = (\alpha_1, \cdots, \alpha_n)$ with $\alpha_i > 0$ and $A_1, \cdots, A_n$, all positive, and writing $(A + \delta)^k = (A_1 + \delta)^{k_1} \cdots (A_n + \delta)^{k_n}$ and $k^{k\alpha} = k_1^{k_1\alpha_1} k_2^{k_2\alpha_2} \cdots k_n^{k_n\alpha_n}$, one denotes by $S_{\alpha,A}(R^n)$, [2], [3], [4], the set of all $C^\infty(R^n)$ functions φ, with each of which are associated constants $C_\sigma = C_\sigma(\varphi)$ such that

$$| x^k \partial^\sigma \varphi(x) | \leqq C_\sigma (A + \delta)^k k^{k\alpha} \tag{3.5}$$

holds throughout R^n for every $\delta > 0$ and all n-tuples k and σ of nonnegative integers. $S_{\alpha,A}$ can also be described as the set of $\varphi \in C^\infty(R^n)$ which with all their derivatives satisfy

$$| \partial^\sigma \varphi(x) | \leqq C_\sigma e^{-c|x|^{1/\alpha}}, \tag{3.6}$$

where C_σ depends on φ and σ and $e^{-c|x|^{1/\alpha}}$ is an abbreviation standing for $\exp(-c_1 | x_1 |^{1/\alpha_1} - \cdots - c_n | x_n |^{1/\alpha_n})$ and

$$c_i = \alpha_i / e A_i^{1/\alpha_i}, \qquad i = 1, \cdots, n. \tag{3.7}$$

Defining $M_p(x)$ by

$$M_p(x) = e^{c(1-1/p)|x|^{1/\alpha}} \tag{3.8}$$

for $p = 0, 1, \cdots$, it follows [2], [4] that

$$\| \varphi \|_p = \sup_{|\sigma| \leqq p} \sup_{x \in R^n} M_p(x) | \partial^\sigma \varphi(x) | \tag{3.9}$$

is finite for each $\varphi \in S_{\alpha,A}(R^n)$ and natural number p.

Let $\mathfrak{B}_F$ be any one of $\mathfrak{D}[-l, l]$, $\mathcal{S}(R^n)$ or $S_{\alpha,A}(R^n)$. The functions $\| \cdot \|_p$, $p = 0, 1, \cdots$, defined on these spaces by (3.1), (3.4), (3.9), respectively, are norms (i.e., $\| \varphi \|_p = 0 \Leftrightarrow \varphi(x) \equiv 0$; for $\lambda \in C$, $\| \lambda\varphi \|_p = | \lambda | \| \varphi \|_p$; and $\| \varphi_1 + \varphi_2 \| \leqq \| \varphi_1 \|_p + \| \varphi_2 \|_p$), which have the further property that

$$\| \varphi \|_0 \leqq \| \varphi \|_1 \leqq \cdots \leqq \| \varphi \|_p \leqq \cdots \tag{3.10}$$

for each $\varphi \in \mathfrak{B}_F$. This is clearly apparent from the fact that in each case one takes a supremum over all σ with $| \sigma | \leqq p$ and that $M_p(x) \leqq M_{p+1}(x)$, whichever of (3.3) or (3.8) defines M_p. Take $\varphi_0 \in \mathfrak{B}_F$, a natural number p and $\epsilon > 0$; the set

$$N_p(\varphi_0 ; \epsilon) = \{\varphi : \varphi \in \mathfrak{B}_F \quad \text{and} \quad \| \varphi - \varphi_0 \|_p < \epsilon\} \tag{3.11}$$

is a spherical neighborhood of φ_0—one might call it an open sphere, or ball, of radius ϵ at the pth level. The collection of such spheres, as p ranges over the nonnegative integers and ϵ over the positive reals, constitutes a so-called fundamental system of neighborhoods of the point φ_0 in $\mathfrak{B}_F$; i.e., interior points, open sets, points of accumulation, closure and all the usual topological concepts are defined in $\mathfrak{B}_F$ by means of these neighborhoods quite as they are in the most elementary accounts of point set topology. In particular, a sequence $\{\varphi_m\} \subset \mathfrak{B}_F$ is said to converge to $\psi \in \mathfrak{B}_F$ (one writes $\lim_m \varphi_m = \psi$) if and only if to every $N_p(\psi; \epsilon)$ corresponds an integer $K = K(p, \epsilon)$ such that $n > K$ implies $\varphi_m \in N_p(\psi; \epsilon)$. Each of these $\mathfrak{B}_F$ has the important property of *completeness*: if $\{\varphi_m\} \subset \mathfrak{B}_F$ is a Cauchy sequence, then there exists ψ such that $\lim_m \varphi_m = \psi$. For $\mathfrak{D}[-l, l]$ and $\mathcal{S}(R^n)$ this is proved in [2], [4], [17] and for $S_{\alpha,A}(R^n)$ in [2], [4]. A *Cauchy sequence* $\{\varphi_m\} \subset \mathfrak{B}_F$ is, by definition, a sequence such that to every $N_p(0; \epsilon)$ there corresponds an integer $M = M(p, \epsilon)$ such that $m, n > M$ imply $(\varphi_n - \varphi_m) \in N_p(0; \epsilon)$ (i.e., $\| \varphi_n - \varphi_m \|_p < \epsilon$). A vector space in which neighborhoods are defined by means of a denumerable collection of norms and which is complete is called an F (for Fréchet) space—our $\mathfrak{B}_F$ are such.

A linear functional on $\mathfrak{B}_F$ is any function[2] $F:\mathfrak{B}_F \to C$ such that for any $\lambda_1, \lambda_2 \in C$ and $\varphi_1, \varphi_2 \in \mathfrak{B}_F$ there holds

$$\langle F, \lambda_1\varphi_1 + \lambda_2\varphi_2 \rangle = \lambda_1\langle F, \varphi_1\rangle + \lambda_2\langle F, \varphi_2\rangle. \tag{3.12}$$

(The notation $\langle F, \varphi\rangle$ for the values of F is preferred to $F(\varphi)$ because its symmetrical appearance is in keeping with the fact that for any two linear functionals F_1, F_2 on $\mathfrak{B}_F$ and any $c_1, c_2 \in C$ the relation $\langle c_1F_1 + c_2F_2, \varphi\rangle = c_1\langle F_1, \varphi\rangle + c_2\langle F_2, \varphi\rangle$ is satisfied for every $\varphi \in \mathfrak{B}_F$.) A function $F:\mathfrak{B}_F \to C$ is continuous at φ_0 if to every $\epsilon > 0$ there corresponds a neighborhood $N_p(\varphi_0; \delta)$ such that $\varphi \in N_p(\varphi_0; \delta)$ implies $| F(\varphi) - F(\varphi_0)| < \epsilon$, i.e., if to every spherical neighborhood $S(F(\varphi_0); \epsilon) = \{z : z \in C \text{ and } | z - F(\varphi_0)| < \epsilon\}$ of $F(\varphi_0)$ in C there corresponds a spherical neighborhood of φ_0 in $\mathfrak{B}_F$ which is mapped by F into a subset of $S(F(\varphi_0); \epsilon)$. In the particular case where F is a linear functional continuous at $0 \in \mathfrak{B}_F$ this means that to $\epsilon = 1$ corresponds a neighborhood $N_p(0; \delta)$ such that $\psi \in N_p(0; \delta)$ implies $|\langle F, \psi\rangle - \langle F, 0\rangle| < 1$; in other terms, $\| \psi \|_p < \delta$ implies $|\langle F, \psi\rangle| < 1$. In this relation take $\psi = \lambda\varphi$, where φ is any element of $\mathfrak{B}_F$ and $\lambda = \delta/(2 \| \varphi \|_p)$, so that $\| \psi \|_p = | \lambda | \| \varphi \|_p = \delta/2 < \delta$; we get $|\langle F, \lambda\varphi\rangle| = | \lambda | |\langle F, \varphi\rangle| < 1$, whence $|\langle F, \varphi\rangle| < (2/\delta)\| \varphi \|_p$. It is essential to note that p does not depend on φ here, nor does δ. Thus with each linear functional F continuous at $0 \in \mathfrak{B}_F$ there is associated a smallest integer $p \geqq 0$, called the *order* of F, such that for some constant B_F there holds

$$|\langle F, \varphi\rangle| \leqq B_F \| \varphi \|_p \tag{3.13}$$

---

[2] The subscript in the symbol $\mathfrak{B}_F$ stands for "Fréchet" and is *in no way* to be identified with the full-sized "F" which designates a functional on $\mathfrak{B}_F$.

for all $\varphi \in \mathfrak{B}_F$. Conversely, if F is linear and satisfies (3.13), then it is obviously continuous at $0 \in \mathfrak{B}_F$. Replacing φ by $\varphi_0 - \varphi$ and using the linearity of F one immediately sees that continuity of a linear functional at $0 \in \mathfrak{B}_F$ implies its continuity at every $\varphi \in \mathfrak{B}_F$. Thus F is a continuous linear functional (c.l.f.) on $\mathfrak{B}_F$ if and only if for some finite p and B_F it satisfies (3.13). The set of all c.l.f. on $\mathfrak{B}_F$ is denoted by $\mathfrak{B}_F'$ and its elements are variously called generalized functions (g.f.) or distributions of class $\mathfrak{B}_F'$.

One more space of functions must be considered here: $\mathfrak{D}(R^n)$, the set of $C^\infty(R^n)$ functions φ, each of which vanishes outside some finite interval (dependent on φ) [2], [3], [4], [6], [13], [17], [19]. This space $\mathfrak{D}(R^n)$ can be represented as

$$\mathfrak{D}(R^n) = \bigcup_{l=2}^{\infty} \mathfrak{D}[-l, l]. \tag{3.14}$$

Besides its obvious set-theoretic meaning one also ascribes to this union a more technical sense in that we define convergence of sequences[3] in $\mathfrak{D}(R^n)$ as follows: $\{\varphi_m\} \subset \mathfrak{D}(R^n)$ converges to $\psi \in \mathfrak{D}(R^n)$ if and only if $\{\varphi_m\} \subset \mathfrak{D}[-l, l]$ for some l and $\lim_m \varphi_m = \psi$ in this same $\mathfrak{D}[-l, l]$. (Since convergence in $\mathfrak{D}[-l, l]$ entails convergence in $\mathfrak{D}(-j, j)$ whenever $l < j$, this definition is consistent.) In this interpretation $\mathfrak{D}(R^n)$ is called the *union space* of the $\mathfrak{D}[-l, l]$. A linear functional on such a union space is continuous if and only if the restriction of F to $\mathfrak{D}[-l, l]$ is a c.l.f. on $\mathfrak{D}[-l, l]$ and this for $j = 1, 2, \cdots$. The set of these F is called the class $\mathfrak{D}'(R^n)$ of L. Schwartz distributions.

Henceforth $\mathfrak{B}$ will denote any of the spaces $\mathfrak{D}(R^n)$, $\mathcal{S}(R^n)$, $S_{\alpha,A}(R^n)$, which are called test spaces, the $\varphi \in \mathfrak{B}$ being known as test functions, while $\mathfrak{B}'$ denotes the collection of c.l.f., i.e., distributions, on $\mathfrak{B}$. $\mathfrak{D}'$ occurs in many applications of distribution theory, $\mathcal{S}'$ and $S'_{\alpha,A}$ are especially useful in applications involving the Fourier transformation [2], [3], [4], [17], [19]. $\mathfrak{B}_F$ will denote $\mathfrak{D}[-l, l]$, $\mathcal{S}(R^n)$ or $S_{\alpha,A}(R^n)$.

4. Test cones. Basic sequences. Let $V(\mathfrak{D})$ be $C(R^n)$, the set of continuous $f: R^n \to C$. Those f in this set whose growth at infinity is restricted by an inequality

$$|f(x)| \leq K_f M_p(x) \tag{4.1}$$

(here K_f and $p = p(f)$ depend on f but not on x) constitute the sets $V(\mathcal{S})$ and $V(S_{\alpha,A})$ respectively, depending on whether M_p is defined by (3.3) or

[3] For our purposes it suffices to consider sequences; more general objects of convergence, e.g., filters, nets, can also be considered, and quite analogously. Reference [13] treats nets in union spaces.

(3.8). It is easy enough to verify that on defining $\langle f, \varphi \rangle$ by

$$\langle f, \varphi \rangle = \int_{R^n} f(x)\varphi(x)\,dx, \tag{4.2}$$

$V(\mathfrak{D})$, $V(\mathfrak{S})$ and $V(S_{\alpha,A})$ become sets of distributions on $\mathfrak{D}$, $\mathfrak{S}$, $S_{\alpha,A}$, respectively.

All the test functions considered here have $\int_{R^n} \varphi(x)\,dx < \infty$. We call $\Pi(\mathfrak{V})$ the subset of $\mathfrak{V}$ consisting of the nonnegative-valued φ, and $\Delta(\mathfrak{V})$ the subset of $\Pi(\mathfrak{V})$ whose elements have $\int_{R^n} \varphi(x)\,dx = 1$, i.e., all probability densities in $\mathfrak{V}$. Neither the $\Delta(\mathfrak{V})$ nor the $\Pi(\mathfrak{V})$ are vector spaces; the latter, though, are convex cones with vertex zero (i.e., $0 \in \Pi(\mathfrak{V})$; and $\lambda > 0$, $\varphi \in \Pi(\mathfrak{V})$ imply $\lambda\varphi \in \Pi(\mathfrak{V})$; also, $(1 - \lambda)\varphi_1 + \lambda\varphi_2 \in \Pi(\mathfrak{V})$ whenever $0 \leqq \lambda \leqq 1$ and $\varphi_1, \varphi_2 \in \Pi(\mathfrak{V})$).

Given $\Delta(\mathfrak{V})$ let us call *basic sequence* any sequence $\{f_\nu\} \subset V(\mathfrak{V})$ such that $\lim_\nu \langle f_\nu, \varphi \rangle$ exists at every $\varphi \in \Delta(\mathfrak{V})$. Two basic sequences $\{f_\nu\}$ and $\{g_\nu\}$ are *equivalent* if $\lim_\nu \langle f_\nu, \varphi \rangle = \lim_\nu \langle g_\nu, \varphi \rangle$ at every $\varphi \in \Delta(\mathfrak{V})$; one then writes $\{f_\nu\} \sim \{g_\nu\}$ and, since the relation has the usual trappings of reflexivity, symmetry and transitivity, obtains a partition of $V_\Sigma(\mathfrak{V})$, the set of basic sequences on $\Delta(\mathfrak{V})$, into disjoint equivalence classes $\mathrm{Cl}(\{f_\nu\}) = \{\{g_\nu\} : \{g_\nu\} \sim \{f_\nu\}\}$. A basic sequence defines a functional $F:\Delta(\mathfrak{V}) \to C$ whose values $\langle F, \varphi \rangle$ are given by

$$\langle F, \varphi \rangle = \lim_\nu \langle f_\nu, \varphi \rangle. \tag{4.3}$$

The collection of these functionals is designated by $\Delta'(\mathfrak{V})$—it is quite obviously in one-to-one correspondence with the collection of equivalence classes whose union is $V_\Sigma(\mathfrak{V})$. In fact, equipping these collections with their natural vector space structures ($F + G$ defined by $\langle F + G, \varphi \rangle = \langle F, \varphi \rangle + \langle G, \varphi \rangle$, cF by $\langle cF, \varphi \rangle = c\langle F, \varphi \rangle$, and $\mathrm{Cl}(\{f_\nu\}) + \mathrm{Cl}(\{g_\nu\})$ and $c\,\mathrm{Cl}(\{f_\nu\})$ by $\mathrm{Cl}(\{f_\nu + g_\nu\})$ and $\mathrm{Cl}(\{cf_\nu\})$, respectively) reveals them to be isomorphic. The burden of §§6, 7 is to prove that $\Delta'(\mathfrak{V}) = \mathfrak{V}'$. In preparation for this observe that if $\lim_\nu \langle f_\nu, \varphi \rangle$ exists for every $\varphi \in \Delta(\mathfrak{V})$ it does so for $\psi \in \Pi(\mathfrak{V})$ as well, since, taking $\int \psi(x)\,dx = v$ and $\varphi = \psi/v$, we have, by (4.2) and (4.3), $\lim_\nu \langle f_\nu, \psi \rangle = v \lim_\nu \langle f_\nu, \varphi \rangle$. Conversely, if $\lim_\nu \langle f_\nu, \varphi \rangle$ exists for every φ in $\Pi(\mathfrak{V})$, it does so for every $\psi \in \Delta(\mathfrak{V})$. It is clear then that in defining basic sequences and the attendant functionals we might just as well have taken the test functions to be in $\Pi(\mathfrak{V})$ and to have denoted the resulting set of $F:\Pi(\mathfrak{V}) \to C$ by $\Pi'(\mathfrak{V})$. This we henceforth do, because we shall find it convenient mathematically—it allows us to write $\langle F, a_1\varphi_1 + a_2\varphi_2\rangle = a_1\langle F, \varphi_1\rangle + a_2\langle F, \varphi_2\rangle$ whenever $a_1, a_2 \geqq 0$. (The point in starting out with $\Delta(\mathfrak{B})$ resides with the physical motivation.)

Remark. Both $\Delta(\mathcal{S})$ and $\Delta(S_{\alpha,A})$ contain numerous Gaussian frequency functions which, because of the central limit theorem, are of the utmost practical significance. As to $\Delta(\mathfrak{D})$, it too contains many physically relevant probability densities. Explicit instances come from [12], whose results can be interpreted in terms of probabilities as showing that many functions in $\mathfrak{D}$ are probability densities of infinite sums of independent random variables with rectangular densities.

Physical investigations which hinge on the explicit use of the probability distribution of independent variable determinations in the observation process are to be found in the literature [5], [7], [8]. The densities considered in [7], [8] are Gaussian, with specified standard deviations corresponding to a "fundamental" length and time below which it is impossible to distinguish the positions of two particles in space or time. Such test functions are of the $\Delta(S_{\alpha,A})$ variety with $\alpha = \frac{1}{2}$ and A determined by the fundamental length and time.

5. Basic lemma. Lemma 2 below is instrumental in later sections. Let $|\cdot|_p:\Pi(\mathfrak{B}_F) \to R$ be the restriction of the norm function $\|\cdot\|_p:\mathfrak{B}_F \to R$ to the domain $\Pi(\mathfrak{B}_F)$ and, with it, define spherical neighborhoods of level p in $\Pi(\mathfrak{B}_F)$ quite as we did in $\mathfrak{B}_F$. Since a Cauchy sequence in $\Pi(\mathfrak{B}_F)$ is Cauchy in $\mathfrak{B}_F$, it converges to some $\varphi_0 \in \mathfrak{B}_F$; but $\varphi_m(x) \geqq 0$ implies $\lim_m \varphi_m(x) \geqq 0$, so every Cauchy sequence in $\Pi(\mathfrak{B}_F)$ converges to some $\varphi_0 \in \Pi(\mathfrak{B}_F)$, which is thus complete. Λ being an arbitrary, not necessarily countable, index set, a family $\{F_\lambda : F_\lambda \in \mathfrak{B}_F', \lambda \in \Lambda\}$ of distributions of class $\mathfrak{B}_F'$ is said to be weakly bounded on $\Pi(\mathfrak{B}_F)$ if for every $\varphi \in \Pi(\mathfrak{B}_F)$ there holds

$$\sup_\lambda |\langle F_\lambda, \varphi\rangle| = S(\varphi) < \infty. \tag{5.1}$$

LEMMA 1. *If the countable family $\{F_\nu\}$, $\nu = 1, 2, \cdots$, of distributions is weakly bounded on $\Pi(\mathfrak{B}_F)$, then there exists a spherical neighborhood $P_p(\varphi_0; \delta)$ in this cone on which $\{F_\nu\}$ is uniformly bounded; i.e., for some $\varphi_0 \in \Pi(\mathfrak{B}_F)$ and $\delta > 0$,*

$$|\langle F_\nu, \varphi\rangle| \leqq K, \tag{5.2}$$

where $K < \infty$ is independent of ν and φ ranges over $P_p(\varphi_0; \delta) = \{\varphi : \varphi \in \Pi(\mathfrak{B}_F)$ and $|\varphi - \varphi_0|_p < \delta\}$.

The slickest proof of this rests on the observation that $\Pi(\mathfrak{B}_F)$ is a topological space of the second category; another employs the Lebesgue resonance procedure. The one we shall give here is an adaptation of an argument originally due to Osgood [16, p. 62]. Preliminary to it we observe that, denoting by $p(\nu)$ the order (see (3.13)) of F_ν, we can always suppose $p(1) \leqq p(2) \leqq \cdots \leqq p(\nu) \leqq \cdots$ since otherwise a relabeling of the F_ν will always achieve this. Also, the corresponding constants B_{F_ν} in (3.13) may be assumed to be monotonically disposed, $B_1 \leqq B_2 \leqq \cdots \leqq B_\nu \leqq \cdots$ because, if they are not, we merely replace the violators by larger constants that do string out in such nondecreasing order. With these dispositions, our family satisfies

$$(5.3) \qquad |\langle F_\nu , \varphi\rangle| \leqq B_\nu \| \varphi \|_{p(\nu)} ,$$

where $\nu = 1, 2, \cdots$. From the linearity of F_ν follows $|\langle F_\nu , \varphi\rangle| = |\langle F_\nu , \psi\rangle + \langle F_\nu , \varphi - \psi\rangle| \geqq |\langle F_\nu , \psi\rangle| - |\langle F_\nu , \varphi - \psi\rangle|$ and, applying (5.3) to the last term,

$$|\langle F_\nu , \varphi\rangle| \geqq |\langle F_\nu , \psi\rangle| - B_\nu \| \varphi - \psi \|_{p(\nu)}$$

for all $\psi, \varphi \in \Pi(\mathfrak{B}_F)$. We have only to take $\| \varphi - \psi \|_{p(\nu)} < 1/B_\nu$ to ensure that

$$(5.4) \qquad |\langle F_\nu , \varphi\rangle| \geqq |\langle F_\nu , \psi\rangle| -1.$$

Thus if $\psi \in \Pi(\mathfrak{B}_F)$ and we confine $\varphi \in \Pi(\mathfrak{B}_F)$ to the sphere $P_{p(\nu)}(\psi; \epsilon)$ with $\epsilon \leqq 1/B_\nu$, (5.4) will hold. In this connection note that, given any sphere $P_p(\psi; \epsilon)$ and any $\theta \in \Pi(\mathfrak{B}_F)$ in this sphere, we have $P_p(\theta; \delta) \subset P_p(\psi; \epsilon)$ providing $\delta < \epsilon - \| \psi - \theta \|_p$; this with (3.10) enables us to assert that

$$(5.5) \qquad P_q(\theta; \delta) \subset P_p(\psi; \epsilon) \qquad \forall \delta \leqq \epsilon - \| \psi - \theta \|_p \quad \text{and} \quad q \geqq p.$$

To prove the lemma suppose the conclusion (5.2) to be false. In that case, $S(\varphi) = \sup_\nu |\langle F, \varphi\rangle|$ is not bounded in any sphere of $\Pi(\mathfrak{B}_F)$; not in $P_1(0, 1)$ among others. But then a $\nu_1 \geqq 1$ and $\varphi_1 \in P_1(0; 1)$ exist such that $|\langle F_{\nu_1} , \varphi_1\rangle| \geqq 2$. Taking $\delta_1 = 1/B_{\nu_1}, q(1) = p(\nu_1)$ and $S_1 = P_{q(1)} (\varphi_1; \delta_1)$, (5.4) indicates that $| \langle F_{\nu_1}, \varphi\rangle | \geqq 1$ for all $\varphi \in S_1$. Now, if the lemma is false, $S(\varphi)$ is not bounded in S_1 either, and so there exist $\nu_2 > \nu_1$ and $\varphi_2 \in S_1$, for which $|\langle F_{\nu_2}, \varphi_2\rangle| \geqq 3$. Set $\delta_2 = \min (1/B_{\nu_2}, \delta_1/2, \delta_1 - \| \varphi_1 - \varphi_2 \|_{q(1)})$ and $q(2) = \max (p(\nu_2), q(1) + 1)$; the sphere $S_2 = P_{q(2)}(\varphi_2 ; \delta_2)$ is by (5.5) contained in S_1, and in view of (5.4), $|\langle F_{\nu_1} , \varphi\rangle| \geqq 1$ and $|\langle F_{\nu_2} , \varphi\rangle| \geqq 2$ hold throughout S_2. Pursuing this routine, with

$$\delta_k = \min (1/B_{\nu_k} , \delta_{k-1}/2, \delta_{k-1} - \| \varphi_{k-1} - \varphi_k \|_{q(k-1)})$$

and $q(k) = \max (p(\nu_k), q(k - 1) + 1)$, we obtain a sequence of spheres $S_k = P_{q(k)}(\varphi_k ; \delta_k)$ at increasing levels $(q(k - 1) < q(k))$ such that $S_k \subset S_{k-1}$ and $|\langle F_{\nu_i} , \varphi\rangle| \geqq i$ for all $i \leqq k$ and $\varphi \in S_k$. But this, since each S_k contains all the S_l with $l > k$ (and therefore their centers φ_l) means that we have

$$(5.6) \qquad |\langle F_{\nu_i}, \varphi_l \rangle| \geqq i \quad \text{whenever} \quad i \leqq l.$$

The sequence $\{\varphi_l\}$ of centers is seen to be a Cauchy sequence and so there exists $\varphi_0 \in \Pi(\mathfrak{B}_F)$ with $\lim_l \varphi_l = \varphi_0$. Letting $l \to \infty$ in (5.6) we arrive at $|\langle F_{\nu_i}, \varphi_0 \rangle| \geqq i$, $i = 1, 2, \cdots$. This contradicts the hypothesis of weak boundedness on $\Pi(\mathfrak{B}_F)$ and proves the lemma.

Along with the sphere $P_p(\varphi_0; \delta)$ in which (5.2) is valid consider the sphere $P_p(0; \delta)$ of the same level p and radius δ and take any $\psi \in P_p(0; \delta) \subset \Pi(\mathfrak{B}_F)$. We then have $(\varphi_0 + \psi) \in P_p(\varphi_0; \delta) \subset \Pi(\mathfrak{B}_F)$ and, since each F_ν is linear,

$$|\langle F_\nu, \psi \rangle| = |\langle F_\nu, \varphi_0 + \psi \rangle - \langle F_\nu, \varphi_0 \rangle|$$
$$\leqq |\langle F_\nu, \varphi_0 + \psi \rangle| + |\langle F_\nu, \varphi_0 \rangle| \leqq K + K,$$

where the last inequality results from (5.2). Thus there exists in $\Pi(\mathfrak{B}_F)$ a neighborhood $P_p(0; \delta)$ of the origin throughout which $|\langle F_\nu, \varphi \rangle|$ is uniformly bounded: $|\langle F_\nu, \psi \rangle| \leqq 2K$. Substituting $\psi = \delta\varphi / |\varphi|_p$ in this inequality and using the linearity of the F_ν we have the proof of the next lemma.

LEMMA 2. *If the family $\{F_\nu\}$ of $\mathfrak{B}_F'$ distributions is weakly bounded on the cone $\Pi(\mathfrak{B}_F)$, then there exist an integer $p \geqq 0$ and a constant B such that*

$$(5.7) \qquad |\langle F_\nu, \varphi \rangle| \leqq B |\varphi|_p$$

throughout this cone. (In other terms: if $\{F_\nu\} \subset \mathfrak{B}_F'$ is weakly bounded on $\Pi(\mathfrak{B}_F)$, then, as a family of functionals on the cone $\Pi(\mathfrak{B}_F)$, $\{F_\nu\}$ is equicontinuous at 0.)

These results remain true if the countable $\{F_\nu\}$ is replaced by an arbitrary parametric family $\{F_\lambda : F_\lambda \in \mathfrak{B}_F', \lambda \in \Lambda\}$. If not, it would always contain a subsequence $\{F_{\lambda_i}\}$, $i = 1, 2, \cdots$, that violates (5.7).

6. Distributions as basic sequences on test cones. Let

$$\rho(x_k) = c \exp(1/(x_k^2 - 1))$$

when $|x_k| < 1$ and $\rho(x_k) = 0$ when $|x_k| \geqq 1$; c is a normalizing constant chosen to make $\int_R \rho(x_k)\, dx_k = 1$. This $C^\infty(R)$ function is nonnegative and has for its support the interval $[-1, 1]$. The function $\rho_\epsilon(x_k) = \rho(x_k/\epsilon)/\epsilon$ has $\operatorname{supp}(\rho_\epsilon) = [-\epsilon, \epsilon]$ and $\int_{-\infty}^{\infty} \rho_\epsilon(x_k)\, dx_k = 1$. Given $x, \xi \in R^n$ we define $\rho_\epsilon(x - \xi) \in \Delta(\mathfrak{D}(R^n)) \subset \Pi(\mathfrak{B}_F)$ by

$$(6.1) \qquad \rho_\epsilon(x - \xi) = \rho_\epsilon(x_1 - \xi_1)\rho_\epsilon(x_2 - \xi_2) \cdots \rho_\epsilon(x_n - \xi_n);$$

its support is the symmetric interval $[\xi_1 - \epsilon, \xi_1 + \epsilon] \times \cdots \times [\xi_n - \epsilon, \xi_n + \epsilon]$ of R^n which we henceforth denote by $[\xi - \epsilon, \xi + \epsilon]$. The $C^\infty(R)$ function

$\int_{-\infty}^{x_k} \rho_{1/2}(\tau_k)\, d\tau_k = \sigma_+(x_k)$ is nonnegative, vanishes for $x_k \leqq -\frac{1}{2}$ and has the constant value 1 for $x_k \geqq \frac{1}{2}$; thus $\sigma_-(x_k) = 1 - \sigma_+(x_k)$ is also nonnegative but vanishes when $x_k \geqq \frac{1}{2}$ and equals 1 for $x_k \leqq -\frac{1}{2}$. The 2^n "octants" of R^n may be described by means of $R_+ = [0, +\infty)$ and $(-1)R_+ = (-\infty, 0]$ as the various Cartesian products of the form $Q_j = (-1)^{i_1}R_+ \times (-1)^{i_2}R_+ \times \cdots \times (-1)^{i_n}R_+$, where each of $i_1, i_2, \cdots, i_n$ takes the value 0 or 1. Let $s(a)$ stand for the translation operator in R: shift by a to the left. Then $s(a)R_+ = [-a, +\infty)$, and the Cartesian product $(-1)^{i_1}s(a)R_+ \times \cdots \times (1)^{i_n}s(a)R_+$ represents a translate of Q_j, which we shall call $Q_j(-a)$ and which if $a > 0$ contains Q_j since it consists of this set plus strips of width a added to its borders (j runs from 1 to 2^n, each j uniquely designating a particular choice of the n-tuple $(i_1, \cdots, i_n)$). With each such $j \leftrightarrow (i_1, \cdots, i_n)$ we associate the function $\alpha_j(x) \in C^\infty(R^n)$ defined as a product

$$\alpha_j(x) = \sigma_\pm(x_1)\sigma_\pm(x_2) \cdots \sigma_\pm(x_n), \tag{6.2}$$

where the subscript chosen for the kth factor is $+$ if $i_k = 0$ and $-$ if $i_k = 1$. This function is nonnegative, carried in the translated octant $Q_j(-\frac{1}{2})$, and is identically equal to one in the translated octant $Q_j(\frac{1}{2}) \subset Q_j$. Note that

$$\sum \alpha_j(x) \equiv 1, \tag{6.3}$$

where by $\sum$ is meant the sum over j from 1 to 2^n.

We now come to the proof that various classes of distributions can be defined as functionals on test functions which are probability distributions. The notation involved was set forth in §4.

THEOREM 1.

$$\Delta'(\mathfrak{D}(R^n)) = \Pi'(\mathfrak{D}(R^n)) = \mathfrak{D}'(R^n).$$

The first equality in the statement is trivial, as was pointed out in §4. Also we know [2, p. 79], [10, p. 193], [19, p. 137] that for every $F \in \mathfrak{D}'$ can be found a sequence $\{f_\nu\}$ of $C(R^n)$—$C^\infty(R^n)$ in fact—functions such that $\langle F, \varphi\rangle = \lim_\nu \langle f_\nu, \varphi\rangle$ at every $\varphi \in \mathfrak{D}(R^n)$, i.e., $\{f_\nu\}$ is a fundamental sequence on $\mathfrak{D}(R^n)$. Since $\Pi(\mathfrak{D})$ is a subset of $\mathfrak{D}$, it follows that $\{f_\nu\}$ is also a basic sequence: $\lim_\nu \langle f_\nu, \varphi\rangle$ exists at every $\varphi \in \Pi(\mathfrak{D})$. Thus the restriction of every $F \in \mathfrak{D}'$ to $\Pi(\mathfrak{D})$ belongs to $\Pi'(\mathfrak{D})$. Furthermore, because two distinct members of $\mathfrak{D}'$ have distinct restrictions to $\Pi(\mathfrak{D})$ (see §7, Remark 2), we have the inclusion $\Pi'(\mathfrak{D}(R^n)) \supset \mathfrak{D}'(R^n)$. There remains to show $\Pi'(\mathfrak{D}) \subset \mathfrak{D}'$. To this end consider an arbitrary sequence $\{f_\nu\}$ of $C(R)$ functions which is basic on $\Pi(\mathfrak{D})$ and therefore on the cone $\Pi(\mathfrak{D}[-l, l]) \subset \Pi(\mathfrak{D})$, where l is any given integer, $l \geqq 2$. There is no loss of generality in supposing these f_ν to be *real*-valued, for the real and imaginary parts of a basic sequence are each basic by themselves. From (6.3) we obtain

(6.4) $$\langle f_\nu , \varphi\rangle = \sum \langle \alpha_j f_\nu , \varphi\rangle.$$

Now, $\alpha_j\varphi \in \Pi(\mathfrak{D}[-l, l])$ whenever $\varphi \in \Pi(\mathfrak{D}[-l, l])$, and $\lim_\nu \langle f_\nu , \alpha_j\varphi\rangle$ consequently exists; since $\langle \alpha_j f_\nu , \varphi\rangle = \int_{R^n} \alpha_j(x) f_\nu(x)\varphi(x)\, dx = \langle f_\nu , \alpha_j\varphi\rangle$, $\lim_\nu \langle \alpha_j f_\nu , \varphi\rangle$ also exists and equals $\langle F, \alpha_j\varphi\rangle$. This convergence entails the weak boundedness on $\Pi(\mathfrak{D}[-l, l])$ of each one of the sequences $\{\alpha_j f_\nu\}_\nu$, and, drawing on Lemma 2, we may assert the existence of constants B_j and integers $p_j \geqq 0$ for which $|\langle \alpha_j f_\nu , \varphi\rangle| \leqq B_j |\varphi|_{p_j}$. Taking $B = \max_j (B_j)$ and $p = \max_j p_j$ it then follows (remember, $|\varphi|_p \leqq |\varphi|_q$ if $p \leqq q$) that, for all $\varphi \in \Pi(\mathfrak{D}[-l, l])$ and $j = 1, 2, \cdots , 2^n$,

(6.5) $$|\langle \alpha_j f_\nu , \varphi\rangle| \leqq B |\varphi|_p .$$

Define $D_k^{-r}(\alpha_j f_\nu)$ by

$$D_k^{-r}(\alpha_j f_\nu) = \int_{\pm\infty}^{x_k} \int_{\pm\infty}^{t_{r-1}} \cdots \int_{\pm\infty}^{t_1} \alpha_j(\cdots x_{k-1}, t_0, x_{k+1}, \cdots)$$

(6.6) $$\cdot f_\nu(\cdots x_{k-1}, t_0, x_{k+1}, \cdots)\, dt_0\, dt_1 \cdots dt_{r-1}$$

$$= \int_{\pm\infty}^{x_k} \frac{(x_k - t)^{r-1}}{(r-1)!} \alpha_j(\cdots , t, \cdots) f_\nu(\cdots , t, \cdots)\, dt,$$

where we choose $-\infty$ as the lower limit of each integral if the kth factor in (6.2) is the "forward" function $\sigma_+(x_k)$ and $+\infty$ if this factor is the "backward" function $\sigma_-(x_k)$. This integral is r times differentiable with respect to x_k with the result $\partial_k^r(D_k^{-r}(\alpha_j f_\nu)) = \alpha_j(x) f_\nu(x)$. Coupled with D_k^{-r} we shall later consider $\bar{D}_k^{-r}$, which differs from D_k^{-r} in that the directions of the integrals in $\bar{D}_k^{-r}$ are the reverse of those in D_k^{-r}. That is, if the integrals in (6.6) are forward (lower limit $-\infty$), those in $\bar{D}_k^{-r}$ are backward (lower limit $+\infty$); and if D_k^{-r} is backward (lower limits in (6.6) $+\infty$), then $\bar{D}_k^{-r}$ is forward (lower limit $-\infty$). Then, from $\int_{R^n} \alpha_j f_\nu \varphi\, dx = \langle \alpha_j f_\nu , \varphi\rangle$, it follows on integrating by parts that $\langle \alpha_j f_\nu , \varphi\rangle = (-1)^r\langle D_k^{-r}(\alpha_j f_\nu), \partial_k^r\varphi\rangle$. For greater brevity put

(6.7) $$D^{-r} = D_1^{-r} D_2^{-r} \cdots D_n^{-r} \quad \text{and} \quad D^r = \partial_1^r \partial_2^r \cdots \partial_n^r.$$

We then obtain

(6.8) $$\langle \alpha_j f_\nu , \varphi\rangle = (-1)^{nr}\langle D^{-r}(\alpha_j f_\nu), D^r\varphi\rangle.$$

Taking $r = p + 1$, where p is determined by (6.5), this yields

(6.9) $$|\langle D^{-p-1}(\alpha_j f_\nu), D^{p+1}\varphi\rangle| \leqq B |\varphi|_p .$$

The continuous function $|D^{-p-1}(\alpha_j f_\nu)|$ has a maximum $\| D^{-p-1}(\alpha_j f_\nu)\|_{[-l,l]}$

in $[-l, l]$. There is consequently a symmetric interval $I_\nu = [\xi - \epsilon, \xi + \epsilon] \subset [-l, l]$ such that, for all $x \in I_\nu$,

$$|D^{-p-1}(\alpha_j(x)f_\nu(x))| \geqq \tfrac{1}{2} \| D^{-p-1}(\alpha_j f_\nu)\|_{[-l,l]} . \tag{6.10}$$

The nonnegative function $\rho_\epsilon(x - \xi)$ of (6.1) is carried in I_ν; therefore,

$$\begin{aligned} |\langle D^{-p-1}(\alpha_j f_\nu), \rho_\epsilon(x - \xi)\rangle| &= \left| \int_{I_\nu} D^{-p-1}(\alpha_j(x)f_\nu(x))\rho_\epsilon(x - \xi)\, dx \right| \\ &\geqq \tfrac{1}{2}\| D^{-p-1}(\alpha_j f_\nu) \|_{[-l,l]} , \end{aligned} \tag{6.11}$$

where the inequality results from (6.10) and $\int_{I_\nu} \rho_\epsilon(x - \xi)\, dx = 1$. Consider the kth factor on the right in (6.2). If it is $\sigma_+(x_k)$, replace it by $\sigma_+(x_k + 1)$; and, if it is $\sigma_-(x_k)$, replace it by $\sigma_-(x_k - 1)$. Doing this for each of the n factors yields a function $\beta_j(x)$ which is a translation of α_j so arranged that $\beta_j(x) = 1$ on supp $(\alpha_j) = Q_j(-\frac{1}{2})$ and which has supp $(\beta_j) = Q_j(-\frac{3}{2})$. Depending on whether the number of backward integrations called for in $\bar{D}^{-p-1}$ is even or odd, the function $\bar{D}^{-p-1}\rho_\epsilon(x - \xi)$ is either nonnegative or nonpositive. Therefore ($\beta_j(x)$ being nonnegative) the function ψ, whose values are

$$\psi(x) = (\pm)\beta_j(x)\bar{D}^{-p-1}\rho_\epsilon(x - \xi) \tag{6.12}$$

(where we choose $(+)$ if $\bar{D}^{-p-1}\rho_\epsilon \geqq 0$ and $(-)$ otherwise) belongs to $\Pi(\mathfrak{D}(R^n))$—to $\Pi(\mathfrak{D}[-l, l])$, actually. This function has, for our purposes, the virtue that it equals $(\pm)\bar{D}^{-p-1}\rho_\epsilon(x - \xi)$ on supp $(\alpha_j f_\nu) \subset$ supp (α_j), whence $\alpha_j(x)f_\nu(x)\psi(x) = \alpha_j(x)f_\nu(x)[(\pm)\bar{D}^{-p-1}\rho_\epsilon(x - \xi)]$ and also

$$D^{p+1}\psi(x) = (\pm)D^{p+1}(\bar{D}^{-p-1}\rho_\epsilon(x - \xi)) = (\pm)\rho_\epsilon(x - \xi)$$

on supp $(\alpha_j f_\nu) \supset$ supp $(D^{-p-1}(\alpha_j f_\nu))$. It is these relations which justify our writing

$$\begin{aligned} \langle \alpha_j f_\nu , \psi\rangle &= (-1)^{n(p+1)}\langle D^{-p-1}(\alpha_j f_\nu), D^{p+1}\psi\rangle \\ &= (-1)^{n(p+1)}\langle D^{-p-1}(\alpha_j f_\nu), \rho_\epsilon(x - \xi)\rangle. \end{aligned} \tag{6.13}$$

Recalling (6.9) and (6.11), we end up with

$$\tfrac{1}{2} \| D^{-p-1}(\alpha_j f_\nu)\|_{[-l,l]} \leqq B\,|\psi|_p . \tag{6.14}$$

This inequality is the crux of the proof of this and the next theorem. From it we shall obtain a bound $M \geqq \| D^{-p-1}(\alpha_j f_\nu)\|_{[-l,l]}$ that is independent of any test function. After that, proving F to be a distribution will be easy. To find this M we have the following lemma.

LEMMA 3. *For every* $\sigma = (\sigma_1, \cdots, \sigma_n)$ *with* $\sigma_1 \leqq p, \cdots, \sigma_n \leqq p$ *the function* ψ *defined in* (6.12) *satisfies the inequalities*

$$\max_x |\partial^\sigma\psi(x)| \leqq A^n 2^{np}(l + 2)^{np}, \tag{6.15}$$

where A is a constant independent of the integer $l = 2, 3, \cdots$ and $\operatorname{supp}(\psi) \subset [-l, l] \subset R^n$ *for all l such that $I_\nu \subset [-l, l]$.*

Proof. ψ is a product of functions of a single variable in which the kth factor is either

$$\begin{aligned} b_+(x_k) &= \sigma_+(x_k + 1)\bar{D}_k^{-p-1}(\rho_\epsilon(x_k - \xi_k)) \\ &= \sigma_+(x_k + 1) \int_{+\infty}^{x_k} \cdots \int_{+\infty}^{t_1} \rho_\epsilon(t_0 - \xi_k)\, dt_0 \cdots dt_p \end{aligned} \tag{6.16}$$

or

$$\begin{aligned} b_-(x_k) &= \sigma_-(x_k - 1)\bar{D}_k^{-p-1}(\rho_\epsilon(x_k - \xi_k)) \\ &= \sigma_-(x_k - 1) \int_{-\infty}^{x_k} \cdots \int_{-\infty}^{t_1} \rho_\epsilon(t_0 - \xi_k)\, dt_0 \cdots dt_p. \end{aligned} \tag{6.16'}$$

Consider the first case; the other is handled quite the same way. First note that altering the limits of integration from $+\infty$ to l changes nothing because $\rho_\epsilon(x_k - \xi_k)$ has its support in $[-l, l] \subset R$; second, note that $\sigma_+(x_k + 1)$ vanishes for $x_k \leqq -\frac{3}{2}$ and so for $x_k \leqq -2$ and thus $-2 \leqq x_k \leqq l$ is all we are interested in. Third, note that for any nonnegative integer $q < p$ we have (recall the second equality in (6.6))

$$\begin{aligned} \partial_k^q \bar{D}_k^{-p-1}(\rho_\epsilon(x_k - \xi_k)) &= \bar{D}_k^{q-p}(\bar{D}_k^{-1}\rho_\epsilon) \\ &= \int_l^{x_k} \frac{(x_k - t)^{p-q-1}}{(p - q - 1)!} \int_l^t \rho_\epsilon(\tau - \xi_k)\, d\tau\, dt, \end{aligned}$$

whence

$$|\partial_k^q \bar{D}_k^{-p-1}(\rho_\epsilon(x_k - \xi_k))| \leqq \frac{(l + 2)^{p-q}}{(p - q)!} \leqq (l + 2)^{p-q} \tag{6.17}$$

since $\max_t \left| \int_l^t \rho_\epsilon(\tau - \xi_k)\, d\tau \right| = 1$. Note that (6.17) also holds for $q = p$. Letting

$$\sup_{r \leqq p} \max_x |\partial_k^r \sigma_+(x_k + 1)| = A,$$

the Leibniz formula then gives, in the light of (6.16) and (6.17),

$$\begin{aligned} \partial_k^{\sigma_k} b_+(x_k) &= \sum_{q=0}^{\sigma_k} \binom{\sigma_k}{q} [\partial_k^q(\bar{D}_k^{-p-1}\rho_\epsilon)][\partial_k^{\sigma_k - q}\sigma_+(x_k + 1)] \\ &\leqq A \sum_{q=0}^{\sigma_k} \binom{\sigma_k}{q} (l + 2)^{p-q} \\ &= A(l + 2)^p \left(\frac{1}{l + 2} + 1\right)^q \leqq A 2^p (l + 2)^p. \end{aligned}$$

The same estimate is obtained for $\partial_k^{\sigma_k} b_-(x_k)$. Feeding this into

$$\partial^\sigma \psi(x) = [\partial_1^{\sigma_1} b_\pm(x_1)] \cdots [\partial_n^{\sigma_n} b_\pm(x_n)]$$

we arrive at (6.15). Returning to the proof of Theorem 1, remark that

$$|\psi|_p \leqq \sup_{|\sigma| \leqq p} \max_x |\partial^\sigma \psi(x)| \leqq A^n 2^{np}(l+2)^{np},$$

which we abbreviate as $C(p; l)$. With (6.14) this provides

$$\| D^{-p-1}(\alpha_j f_\nu) \|_{[-l,l]} \leqq 2BC(p; l), \tag{6.18}$$

an inequality which is valid for $\nu = 1, 2, \cdots, j = 1, 2, \cdots, 2^n$, the integer p associated with the sequence $\{f_\nu\}$ and the interval $[-l, l]$ in accord with (6.5).

From (6.18) and the fact that $D^{-p-2}(\alpha_j f_\nu)$ is obtained from $D^{-p-1}(\alpha_j f_\nu)$, by integrating once with respect to each x_k one easily deduces that for each j the sequence, in ν, of the functions $D^{-p-2}(\alpha_j f_\nu)$ is bounded and equicontinuous on $[-l, l] \subset R^n$. Invoking the classical theorem of Arzela-Ascoli we may assert that every subsequence of the sequence itself contains a subsequence which is uniformly convergent on $[-l, l]$ to a continuous function. We now make the claim that these all converge to the same function $G_j: [-l, l] \to C$. Indeed, if it were not so, at least two distinct sequences $\{\nu'(i)\}_i$ and $\{\nu''(k)\}_k$ of integers would exist such that $\lim_{i\to\infty} D^{-p-2}(\alpha_j f_{\nu'(i)}) = A(x)$, $\lim_{k\to\infty} D^{-p-2}(\alpha_j f_{\nu''(k)}) = B(x)$, where A and B are continuous on $[-l, l]$, the convergence is uniform on that interval and $|A(\xi_0) - B(\xi_0)| = d > 0$ at some $\xi_0 \in [-l, l]$. Let $[\xi - \epsilon, \xi + \epsilon]$ be an interval throughout which $|A(x) - B(x)| \geqq 2d/3$. Because of uniform convergence, there exist integers $I > 0$ and $K > 0$ such that, in $[\xi - \epsilon, \xi + \epsilon]$,

$$i > I \Rightarrow |A(x) - D^{-p-2}(\alpha_j(x) f_{\nu'(i)}(x))| < \frac{d}{6},$$

$$k > K \Rightarrow |B(x) - D^{-p-2}(\alpha_j(x) f_{\nu''(k)}(x))| < \frac{d}{6}.$$

It follows that for these i and j, and throughout $[\xi - \epsilon, \xi + \epsilon]$,

$$|D^{-p-2}(\alpha_j f_{\nu'(i)}) - D^{-p-2}(\alpha_j f_{\nu''(k)})| \geqq \frac{2d}{3} - \frac{d}{6} - \frac{d}{6} = \frac{d}{3} > 0. \tag{6.19}$$

Now take $\varphi(x) = (\pm)\beta_j(x)\bar{D}^{-p-2}\rho_\epsilon(x - \xi)$. It is nonnegative for the right choice of $+$ or $-$ and equals $(\pm)\,\bar{D}^{-p-2}\rho_\epsilon(x - \xi)$ on supp (α_j), while $D^{p+2}\varphi = \rho_\epsilon(x - \xi)$ there. It then follows from (6.19) that

$$\lim_{i\to\infty} \langle \alpha_j f_{\nu'(i)}, \varphi \rangle = (-1)^{n(p+2)} \lim_i \langle D^{-p-2}(\alpha_j f_{\nu'(i)}), D^{p+2}\varphi \rangle$$

must differ from

$$\lim_{k\to\infty} \langle \alpha_j f_{\nu''(k)}, \varphi\rangle = (-1)^{n(p+2)} \lim_k \langle D^{-p-2}(\alpha_j f_{\nu''(k)}), D^{p+2}\varphi\rangle.$$

This, however, contradicts the fact that $\{\alpha_j f_\nu\}$ is a basic sequence. Our claim is thereby proved: to each $j = 1, \cdots, 2^n$ corresponds a function $G_j:[-l, l] \to C$ which is continuous and the uniform limit of $\{D^{-p-2}(\alpha_j f_\nu)\}_\nu$. This result we single out for formal mention in Lemma 4.

LEMMA 4. *Given a sequence $\{f_\nu\}$ of $C(R^n)$ functions that is basic on $\Pi(\mathfrak{D}(R^n))$ there corresponds to each interval $[-l, l]$ of R^n an integer $r \geqq 0$ such that, for each of the 2^n functions α_j defined above, the sequence $\{D^{-r}(\alpha_j f_\nu)\}_\nu$ converges uniformly to a continuous function $G_j:[-l, l] \to C$. D^{-r} is defined in (6.6) and (6.7).*

Note that, in the above, r is stated to be nonnegative, not $\geqq 2$ as the preceding proof indicates. This is not an error but simply means that while $r = p + 2$ is sure to work, it may well be that a smaller r will do in some cases.

Given any $\varphi \in \mathfrak{D}[-l, l]$, not necessarily in $\Pi(\mathfrak{D}[-l, l])$, $\langle f_\nu, \varphi\rangle$ is well defined and satisfies (6.4) and (6.8), that is,

$$(6.20) \qquad \langle f_\nu, \varphi\rangle = \sum_j \langle \alpha_j f_\nu, \varphi\rangle = \sum_j (-1)^{nr}\langle D^{-r}(\alpha_j f_\nu), D^r\varphi\rangle,$$

where the r in the last equality is the one provided by Lemma 4. So long as all we knew about $\{f_\nu\}$ was that it was basic, we could meaningfully let ν in (6.20) go to infinity only for $\varphi \in \Pi(\mathfrak{D}[-l, l])$. Now, however, thanks to Lemma 4, this limit makes sense with any $\varphi \in \mathfrak{D}[-l, l]$; for the simple reason that the uniform convergence on $[-l, l]$ of $\{D^{-r}(\alpha_j f_\nu)\}_\nu$ entails

$$\lim_{\nu\to\infty} \int_{R^n} D^{-r}(\alpha_j f_\nu) D^r\varphi \, dx = \int_{[-l,l]} G_j(x) D^r\varphi(x)\, dx = \langle G_j, D^r\varphi\rangle$$

since $D^r\varphi(x)$ is continuous. Each term in the sums in (6.20) is thus convergent, and we conclude that

$$(6.21) \qquad \lim_{\nu\to\infty} \langle f_\nu, \varphi\rangle = (-1)^{nr}\langle \sum_j G_j, D^r\varphi\rangle.$$

The continuous function $G:[-l, l] \to C$ defined by $G = \sum G_j$ has $\gamma = \max_{x\in[-l,l]} |G(x)| < \infty$, while $\max_x |D^r\varphi(x)| = \max_x |\partial_1^r \cdots \partial_n^r\varphi(x)|$ $\leqq \|\varphi\|_{nr}$ in view of the definition (3.1). Consequently, $\int G(x)D^r\varphi(x)\, dx$ $\leqq (2l)^n\gamma \|\varphi\|_{nr}$, which applied on the right of (6.20) yields

$$(6.22) \qquad |\lim_{\nu\to\infty} \langle f_\nu, \varphi\rangle| \leqq (2l)^n \gamma \|\varphi\|_{nr}.$$

The functional $F:\mathfrak{D}[-l, l] \to C$, whose values are $\langle F, \varphi\rangle = \lim \langle f_\nu, \varphi\rangle$, is thus an element of $\mathfrak{D}'[-l, l]$ by (3.13). As this holds for $l = 2, 3, \cdots$

(albeit with different r for different l in most cases), $F \in \mathfrak{D}'(R^n)$. The proof of Theorem 1 is complete.

COROLLARY 1. *Every* $F \in \Pi'(\mathfrak{D}(R^n)) = \mathfrak{D}'(R^n)$ *coincides on every finite open interval* $(a, b) \subset R^n$ *with a distributional derivative of finite order of a function continuous on a closed interval* $[-l, l] \supset (a, b)$.

Proof. ($F_1, F_2 \in \mathfrak{D}'$ are said to coincide on (a, b) if and only if $\langle F_1, \varphi\rangle = \langle F_2, \varphi\rangle$ for every $\varphi \in \mathfrak{D}(R^n)$ having supp $(\varphi) \subset (a, b)$.) Let $F \in \Pi'(\mathfrak{D}(R^n))$ and take any interval of the form $[-l, l]$ containing (a, b). If G is the function associated, as above, with a basic sequence $\{f_\nu\}$, defining F then, by (6.21), $\langle F, \varphi\rangle = (-1)^{nr}\langle G, D^r\varphi\rangle$ for all $\varphi \in \mathfrak{D}[-l, l]$, in particular if supp $(\varphi) \subset (a, b)$. When the distributional derivative $D^rG = \partial_1^r \cdots \partial_n^r G$ is defined by $\langle D^rG, \varphi\rangle = (-1)^{nr}\langle G, D^r\varphi\rangle$, then F coincides with D^rG on (a, b). This result is a form of the fundamental structure theorem of distribution theory and is often stated thus: *locally* every $\mathfrak{D}'(R^n)$ distribution is a finite order derivative (distributional) of a continuous function.

Remark 1. Consider $\{f_\nu\}$, basic on $\Pi(\mathfrak{D}(R^n))$, and $r \geqq 0$ such that $\{D^{-r}(\alpha_j f_\nu)\}_\nu$ converges uniformly on $[-l, l]$. Next, let $\{g_\nu\} \sim \{f_\nu\}$. Does it follow that $\{D^{-s}(\alpha_j g_\nu)\}_\nu$ is also uniformly convergent on $[-l, l]$ when $s = r$? The answer is, no. Lemma 4 provides that for *some* $s \geqq 0$ this is true, but not that $s = r$.

Example. $f_\nu(x) = 0$, $g_\nu(x) = \nu^3 \sin \nu x$, $x \in R$. Then $\{f_\nu(x)\}$ is uniformly convergent on $[-l, l]$ but $\{\alpha_j \nu^3 \sin \nu x\}$ is not; $\{D^{-4}(\alpha_j(x)\nu^3 \sin \nu x)\}_\nu$, however, is. Both define the identically zero functional. On this point see also §7, Remark 2.

Remark 2. Theorem 1, its satellite Lemma 4 and Corollary 1 can be proved directly by classical analysis, without drawing on the language of topology at all. This may be seen by confining the reasoning of [10, p. 188] to non-negative φ and, at the same time, couching it in n dimensions by means of the $D^{-r}(\alpha_j f_\nu)$ device described above. The choice made here—to proceed instead via a little elementary topology and Lemmas 1 and 2—has the advantage of gathering in one place (Lemma 2) structural features that are common to the $\Pi'(\mathfrak{B}_F) \leftrightarrow \mathfrak{B}_F'$ relationship for a very wide class of test spaces of which the three treated here are only a sample. Also, it simplifies the discussion of completeness and relative compactness given below (§8).

7. The $\Pi'(\mathcal{S}(R^n))$ and $\Pi(S'_{\alpha,A}(R^n))$ cases. The proofs of the $\mathcal{S}(R^n)$ and $S_{\alpha,A}(R^n)$ versions of the preceding theorem are obtained from the arguments of §6 by modifications which are sufficiently direct to allow slightly abbreviated description, following the enunciation of these next two theorems.

THEOREM 2.

$$\Delta'(\mathcal{S}(R^n)) = \Pi'(\mathcal{S}(R^n)) = \mathcal{S}'(R^n).$$

THEOREM 3.

$$\Delta'(S_{\alpha,A}(R^n)) = \Pi'(S_{\alpha,A}(R^n) = S'_{\alpha,A}(R^n).$$

As in Theorem 1 the $\Delta' = \Pi'$ statements are trivial, while the truth of $\Pi'(\mathcal{S}) \supset \mathcal{S}'$, $\Pi'(S_{\alpha,A}) \supset S'_{\alpha,A}$ follows, as there, from an argument we temporarily defer (see §7, Remark 2) and from the fact that if $F \in \mathcal{S}'$ or $S'_{\alpha,A}$ then the sequences $\{f_\nu\}$ whose elements are the $C^\infty(R)$ functions $f_\nu(x) = \langle F(t), \rho_{1/\nu}(x - t)\rangle$ are fundamental sequences[4] for F, i.e., $\langle F, \varphi\rangle = \lim_\nu \langle f_\nu , \varphi\rangle$ for every φ in $\mathcal{S}(R^n)$ or $S_{\alpha,A}(R^n)$ respectively. For $F \in \mathcal{S}'$ this is proved in [17, vol. II, p. 95], where the $\{f_\nu\}$ are called *regularizations* of F; for $F \in S'_{\alpha,A}$ a carbon copy of Schwartz's demonstration yields the result. Proofs along the lines of [10, p. 194] can also be given.

Thus we are left with the requirement of proving that

$$\Pi'(\mathfrak{B}_F) \subset \mathfrak{B}_F', \tag{7.1}$$

where $\mathfrak{B}_F$ is $\mathcal{S}(R^n)$ or $S_{\alpha,A}(R^n)$. In other words, given any sequence $\{f_\nu\}$ of $V(\mathfrak{B}_F)$ functions satisfying (4.1) and for which $\lim_\nu \langle f_\nu , \varphi\rangle = \langle F, \varphi\rangle$ exists at every $\varphi \in \Pi(\mathfrak{B}_F)$, show that F is a c.l.f. on $\mathfrak{B}_F$. As before we designate the restriction $\|\cdot\|_p$ to $\Pi(\mathfrak{B}_F)$ by $|\cdot|_p$. This time, though,

$$|\varphi|_p = \sup_{|\sigma| \leqq p} \sup_x M_p(x)|\partial^\sigma\varphi(x)|, \qquad p = 0, 1, \cdots, \tag{7.2}$$

with supp $(\varphi) \subset R^n$ and M_p defined by either (3.3) or (3.8) and an integer $p \geqq 0$. The very same functions α_j, β_j, ψ, and operators D^{-r}, $\bar{D}^{-r}$, D^r encountered in §6 will be used here again.

Lemma 2 is applicable to the present situation, and we find, as in §6, that, for some $B < \infty$ and integer $p \geqq 0$ and all $\varphi \in \Pi(\mathfrak{B}_F)$,

$$|\langle \alpha_j f_\nu , \varphi\rangle| \leqq B|\varphi|_p \tag{7.3}$$

for $j = 1, \cdots, 2^n$, $\nu = 1, 2, \cdots$. From this we get the analogue of (6.9):

$$|\langle D^{-p-1}(\alpha_j f_\nu), D^{p+1}\varphi\rangle| \leqq B|\varphi|_p, \tag{7.4}$$

valid for all $\varphi \in \Pi(\mathfrak{B}_F)$, but with $|\varphi|_p$ as in (7.2).

Next, remark that there is no harm in assuming f_ν to be real-valued since the reasoning which follows can always be applied to $\{g_\nu\}$ and $\{h_\nu\}$ separately if $f_\nu = g_\nu + ih_\nu$. Also, remark that whenever $\varphi \in \Pi(\mathfrak{D}[-l, l])$ then φ is in $\Pi(\mathcal{S})$ and $\Pi(S_{\alpha,A})$ as well, and, therefore, $\{f_\nu\}$ basic on $\Pi(\mathcal{S})$ or $\Pi(S_{\alpha,A})$ implies $\{f_\nu\}$ is basic on $\Pi(\mathfrak{D}[-l, l])$. Thus (3.11) and (3.13) are valid here, too, so that applying them along with (7.4) we find

$$\| D^{-p-1}(\alpha_j f_\nu)\|_{[-l,l]} \leqq B|\psi|_p. \tag{7.5}$$

---

[4] The purpose of the notation $F(t)$ is to indicate that the g.f. F is being considered as a functional on test functions of the variable $t \in R^n$; thus x in $\langle\cdots\rangle$ is a parameter.

Recalling (3.15) and observing that M_p, whether defined by (3.3) or (3.8), is an even and increasing function in each of its n variables x_k, gives us the estimate

$$(7.6) \qquad |\psi|_p \leqq A^n 2^{np}(l+2)^{np} M_p(\bar{l}),$$

where $\bar{l} = (l, l, \cdots, l) \in R^n$; this combined with (7.5) furnishes, for $l = 2, 3, \cdots,$

$$(7.7) \qquad \| D^{-p-1}(\alpha_j f_\nu) \|_{[-l,l]} \leqq E(l+2)^{np} M_p(\bar{l}),$$

with E a constant independent of l. This same estimate applies to the set $T(l) = [-l, l] - [-(l-1), l-1]$ lying outside one interval and inside the next:

$$(7.8) \qquad \| D^{-p-1}(\alpha_j f_\nu) \|_{T(l)} \leqq E(l+2)^{np} M_p(\bar{l}).$$

Now, $x \in T(l)$ implies $l - 1 \leqq |x_k| \leqq l$, $k = 1, \cdots, n$, and so $l \leqq [|x_k|] + 2$ for all $x \in T(l)$. ($[\cdot]$ here is the usual "integer portion" function on R.) Thus, if we let $[x] + 2$ stand for the n-tuple $([|x_1|] + 2, \cdots, [|x_n|] + 2)$, (7.8), whose right-hand side is not decreased if l is replaced by a larger quantity, gives

$$(7.9) \qquad \begin{aligned} &|D^{-p-1}(\alpha_j(x) f_\nu(x))| \\ &\qquad \leqq E([|x_1|] + 4)^p \cdots ([|x_n|] + 4)^p M_p([x] + 2) \end{aligned}$$

for all $x \in R^n$.

When M_p is given by (3.3), the right-hand side of (7.9) gives

$$\begin{aligned} E \prod_{k=1}^{n} ([|x_k|] + 4)^p([|x_k|] + 2)^p &\leqq E \prod_{k=1}^{n} (|x_k| + 4)^{2p} \\ &\leqq c^n E \prod_{k=1}^{n} (|x_k| + 1)^{2p}, \end{aligned}$$

where $c = \sup_{x_k} ((|x_k| + 4)/(|x_k| + 1))^{2p}$. When M_p is given by (3.8), some elementary computation shows that there exist $p' > p$ and $c < \infty$ such that

$$\prod_{k=1}^{n} ([|x_k|] + 4) M_p([x] + 2) \leqq c^n M_{p'}(x).$$

Thus in both cases we deduce that there exist an integer $p' > p$ and a constant $C = c^n E$ such that the right-hand side of (7.9) is majorized by $CM_{p'}(x)$, i.e., (7.8) may be replaced with

$$(7.10) \qquad |D^{-p-1}(\alpha_j(x) f_\nu(x))| \leqq CM_{p'}(x).$$

As the left-hand side vanishes outside supp (α_j) and as $\beta_j = 1$ on supp (α_j), it is permissible to replace the right-hand side of (7.10) by $C\beta_j(x)M_{p'}(x)$ and to apply the operator D^{-1} on both sides of the resulting inequality, whence (since $|D^{-1}g| \leqq D^{-1}|g|$)

$$(7.11) \qquad |D^{-p-2}(\alpha_j(x)f_\nu(x))| \leqq |CD^{-1}\beta_j M_{p'}(x)| \leqq C'M_{p''}(x)$$

for some $p'' > p'$. (The last inequality on the right, here, comes from the observation that, since

$$|D_k^{-1}\beta_j M_{p'}(x)| \leqq \left| \int_{\pm 3/2}^{x_k} M_{p'}(\cdots, t, \cdots)\, dt \right| \leqq \left(|x_k| + \frac{3}{2} \right) M_{p'}(x),$$

we have

$$|D^{-1}\beta_j M_{p'}(x)| \leqq \prod_{k=1}^{n} (|x_k| + 2) M_{p'}(x);$$

and this is majorized by $cM_{p''}(x)$ for some appropriate $p'' > p'$—as indicated just prior to (7.10)). Equations (7.10) and (7.11) make it plain that in *every* bounded interval, $[-l, l]$, for instance, $\{D^{-p-2}(\alpha_j f_\nu)\}_\nu$ is a sequence of equicontinuous functions. But then the argument preceding Lemma 4 is applicable and we conclude that $\{D^{-p-2}(\alpha_j f_\nu)\}_\nu$ converges uniformly on $[-l, l]$. This time however, the integer p does *not* vary with l and, because (7.11) prevails for all $x \in R^n$, we have this next lemma.

LEMMA 5. *Given a sequence $\{f_\nu\}$ of $V(\mathfrak{B}_F)$ functions that is basic on $\Pi(\mathfrak{B}_F)$, where $\mathfrak{B}_F$ is $\mathcal{S}(R^n)$ or $S_{\alpha,A}(R^n)$, respectively, there exist integers $r \geqq 0$ and $s \geqq r$ such that for each of the 2^n functions α_j the sequence $\{D^{-r}(\alpha_j f_\nu)\}_\nu$ converges uniformly on every finite interval of R^n to a continuous function $G_j: R^n \to C$, whose growth at infinity is restricted by*

$$(7.12) \qquad |G_j(x)| \leqq CM_s(x).$$

A similar estimate holds for each function in the sequence

$$(7.13) \qquad |D^{-r}(\alpha_j(x)f_\nu(x))| \leqq CM_s(x).$$

(Note that as in §6 this result is stated in terms of $r \geqq 0$ and $s \geqq r$, not the $r = p + 2$ and the $p'' \geqq p$ of the proof above, because this proof does not necessarily yield the smallest r and s applicable.)

The two types of function $M_p(x)$ considered here both have the property that given s there exists an integer $t > s$ such that

$$\int_{R^n} \frac{M_s(x)}{M_t(x)}\, dx = K < \infty.$$

(In the $\mathcal{S}(R^n)$ case, $t = s + 2$ will do; in the $S_{\alpha,A}$ case, $t = s + 1$ suffices.) Take any $\varphi \in \mathfrak{B}_F$, not necessarily in $\Pi(\mathfrak{B}_F)$, and let $r \leqq s$ be as in Lemma 5 and $t > s$ as just indicated. Then, given $\epsilon > 0$, consider

$$(7.14)\quad \begin{aligned}\langle G_j - D^{-r}(\alpha_j f_\nu), D^r \varphi\rangle &= \int_{R^n} [G_j(x) - D^{-r}(\alpha_j(x) f_\nu(x))] D^r\varphi(x)\, dx \\ &= \left(\int_I + \int_{R^n - I}\right) [G_j - D^{-r}(\alpha_j f_\nu)]\, D^r\varphi(x)\, dx = J_1 + J_2,\end{aligned}$$

where I is a bounded interval in R^n so chosen that

$$\int_{R^n - I} \frac{M_s(x)}{M_t(x)}\, dx = \frac{\epsilon}{2C \max M_t(x) \mid D^r\varphi(x) \mid}.$$

As the integrand of J_1 converges uniformly on I, by Lemma 5, there is an integer N such that $n > N$ implies $J_1 < \epsilon/2$. For J_2 we also have, in view of (7.12), (7.13) and the choice of I,

$$J_2 \leqq 2C \left| \int_{R^n - I} \frac{M_s}{M_t} (M_t\, D^r\varphi)\, dx \right| \leqq \frac{\epsilon}{2}.$$

This proves that

$$\lim_\nu \langle \alpha_j f_\nu, \varphi\rangle = \lim_\nu (-1)^{nr} \langle D^{-r}(\alpha_j f_\nu), D^r\varphi\rangle = (-1)^{nr}\langle G_j, D^r\varphi\rangle$$

for every $\varphi \in \mathfrak{B}_F$. Repeating the computation (6.20) we may assert that $\lim \langle f_\nu, \varphi\rangle$ exists and equals

$$(7.15)\qquad \lim_\nu \langle f_\nu, \varphi\rangle = (-1)^{nr}\langle G, D^r\varphi\rangle,$$

with $G = \Sigma G_j$ a continuous function on R^n whose growth is limited by

$$(7.16)\qquad | G(x) | \leqq 2^n C M_s(x).$$

Thus $F: \Pi(\mathfrak{B}_F) \to C$ is also a linear functional on $\mathfrak{B}_F$ to C with values $\langle F, \varphi\rangle = \lim_\nu \langle f_\nu, \varphi\rangle$ and is a continuous one besides, as we deduce from (7.15) in this manner:

$$\begin{aligned}| \langle F, \varphi\rangle | = \left| \int_{R^n} G\, D^r\varphi\, dx \right| &\leqq 2^n C \int_{R^n} \frac{M_s}{M_t} M_t \mid D^r\varphi \mid dx \leqq 2^n CK \max_x M_t \mid D^r\varphi \mid \\ &\leqq 2^n CK \sup_{|\sigma| \leqq nt} \max_x M_{nt}(x) \mid \partial^\sigma\varphi(x) \mid \\ &= 2^n CK \, \|\varphi\|_{nt}.\end{aligned}$$

Theorems 2 and 3 are thereby proved.

In the course of this reasoning we have also proved, in view of (7.15), these structural properties.

COROLLARY 2. *Every $F \in \Pi'(\mathcal{S}(R^n)) = \mathcal{S}'(R^n)$ is a finite order distributional derivative of the form $F = D^r G$, where the continuous $G: R^n \to C$ is such* *that, for some integer* $s \geqq 0$ *and constant* C,

$$|G(x)| \leqq C \prod_{k=1}^{n} (1 + |x_k|)^s.$$

COROLLARY 3. *Every* $F \in \Pi'(S_{\alpha,A}(R^n)) = S_{\alpha,A}(R^n)$ *is a finite order distributional derivative of the form* $F = D^r G$, *where* $G: R^n \to C$ *is continuous and for some integer* $s \geqq 1$ *and constant* C,

$$|G(x)| \leqq C \exp\left\{\left(1 - \frac{1}{s}\right)(c_1 |x_1|^{1/\alpha_1} + \cdots + c_n |x_n|^{1/\alpha_n})\right\}$$

(the c_i and α_i are described in §3).

Briefly, every $F \in \Pi'(\mathfrak{B}_F) = \mathfrak{B}_F'$ is a finite order distributional derivative of a function in $V(\mathfrak{B}_F)$. That every $F \in \mathcal{S}'(R^n)$ has the structure indicated in Corollary 2 is well known [17, vol. II].

Remark 1. There are other spaces like the two $\mathfrak{B}_F$ considered here for which basic sequences of $V(\mathfrak{B}_F)$ functions determine linear functionals whose ensemble, $\Pi'(\mathfrak{B}_F)$, coincides with the corresponding distribution class $\mathfrak{B}_F'$. Among them are the classes $\mathcal{L}_{c,d}$ and $\mathcal{W}_{a,b}$ defined in [20], [21] and designed for the extension to distributions of certain convolution transforms.

Remark 2. Three different situations where $\Pi'(\mathfrak{B}) = \mathfrak{B}'$ have been described in some detail. Each one of them raises a question that has not been answered so far: does testing with $\Pi(\mathfrak{B})$ functions alone suffice to distinguish between different elements of $\mathfrak{B}'$? Might it not be that, though F_1 and F_2 are distinct members of $\mathfrak{B}'$ (i.e., $\langle F_1 - F_2, \varphi\rangle \neq 0$ for some $\varphi \in \mathfrak{B}$), the test cone $\Pi(\mathfrak{B})$ is not refined enough to perceive this distinction (i.e., $\langle F_1 - F_2, \varphi\rangle = 0$ for all $\varphi \in \Pi(\mathfrak{B})$)? The answer is, no: if $\langle F_1 - F_2, \varphi\rangle = 0$ throughout $\Pi(\mathfrak{B})$, then this equality is also true for all $\varphi \in \mathfrak{B}$. Physically, this is most reassuring: mathematically distinct objects F_1 and F_2 in $\mathfrak{B}'$ are also experimentally distinguishable by means of the testing set $\Pi(\mathfrak{B})$.

The proof is straightforward. We must show that if two basic sequences $\{f_\nu\}, \{g_\nu\}$ on $\Pi(\mathfrak{B})$ are equivalent they define the same $F \in \mathfrak{B}'$. But this, for $\mathfrak{B} = \mathfrak{D}(R^n)$, $\mathcal{S}(R^n)$ or $S_{\alpha,A}(R^n)$, follows from the proofs of the theorems of §6 and §7. Indeed, if $\lim \langle f_\nu - g_\nu, \varphi\rangle = 0$ at every $\varphi \in \Pi(\mathfrak{B})$, then given any interval $[-l, l] \subset R^n$ there exists $r \geqq 0$ such that, as $\nu \to \infty$, $D^{-r}(\alpha_j(f_\nu - g_\nu))$ converges uniformly to $G_j(x)$, continuous on $[-l, l]$. If G_j were not identically zero, an appropriate $\psi \in \Pi(\mathfrak{D}[-l, l])$ (and thus in $\mathcal{S}$ and $S_{\alpha,A}$ as well) of the type (6.12) could be found for which

$$\lim_\nu \langle f_\nu - g_\nu, \alpha_j\psi\rangle = \lim_\nu \langle \alpha_j(f_\nu - g_\nu), \psi\rangle \neq 0.$$

(There is no need to detail this point further here, as the argument is essentially identical to that centering on (6.19).) As $\alpha_j\psi \in \Pi(\mathfrak{B})$ it would follow that $\{f_\nu\}$ is not equivalent to $\{g_\nu\}$ on $\Pi(\mathfrak{B})$, and this contradiction to the original assumption proves our thesis. Thus, not only is $\Pi'(\mathfrak{B})$ a subset of $\mathfrak{B}'$, but the imbedding is one-to-one.

Remark 3. The space $\mathfrak{D}(R^n)$ was defined as a union of $\mathfrak{D}[-l, l]$. A similar process can be applied to spaces S_{α,A_i}, where $A_i < A_{i+1}$ and $\{A_i\}_i$ is unbounded. For $A_i = i$, $i = 1, 2, \cdots$, this has been done in [2], [4]. The distribution space S_α' is then the set of linear functionals on $S_\alpha = \bigcup_{i=1}^\infty S_{\alpha,A_i}$ which are continuous on each individual S_{α,A_i}. Applying the reasoning of §6 to the $\Pi'(S_{\alpha,A_i}) = S'_{\alpha,A_i}$, one concludes that $\Pi'(S_\alpha) = S_\alpha'$. (Naturally, the basic sequences here must be such that $\langle f_\nu, \varphi\rangle$ exists for every $\varphi \in S_\alpha$; this can be achieved by taking $V(S_\alpha)$ to be $\bigcap_{i=1}^\infty V(S_{\alpha,A_i})$. In connection with the fundamental length and time application mentioned in §4, Remark, it might be relevant to replace $\{A_i\} = \{i\}$ by an unbounded increasing sequence $\{A_i\}$ with $\inf A_i \geqq \gamma > 0$. The corresponding $\Pi'(\bigcup_{i=1}^\infty S_{\alpha,A_i}) = (\bigcup_{i=1}^\infty S_{\alpha,A_i})'$ would then describe functionals experimentally observed with testing functions whose "spread" (the usual standard deviation in the case of Gaussian φ) is bounded below by the constant γ related to the fundamental length and time intervals.

8. Completeness and relative compactness in $\Pi'(\mathfrak{B})$. Let us define the *restricted order* of $F \in \mathfrak{B}_F' = \Pi'(\mathfrak{B}_F)$ to be the smallest integer $q \geqq 0$ for which there exists a constant $K_q < \infty$ such that

$$|\langle F, \varphi\rangle| \leqq K_q |\varphi|_q \tag{8.1}$$

for all $\varphi \in \Pi(\mathfrak{B}_F)$.

LEMMA 6. *If $F \in \Pi'(\mathfrak{D}(R^n))$ has, as a member of $\Pi'(\mathfrak{D}[-l-1, l+1])$, restricted order q, then there exist a continuous function $F_{-q}: [-l, l] \to C$ and a constant $A(q, l)$ depending on q and l only (not on F) such that*

$$\| F_{-q}(x)\|_{[-l,l]} \leqq A(q, l)K_q \tag{8.2}$$

and

$$\langle F, \varphi\rangle = (-1)^{(q+3)n}\langle F_{-q}, D^{q+3}\varphi\rangle \tag{8.3}$$

whenever $\varphi \in \mathfrak{D}[-l, l]$.

Proof. We know from Corollary 1 that there exist a continuous function $G: [-(l+1), l+1] \to C$ and an integer $r \geqq 0$ such that F coincides with D^rG on the open interval $(-(l+1), l+1)$. Take any $\varphi \in \Pi(\mathfrak{D}[-l, l])$ and $\rho_{1/(2\nu)}(x)$ as defined in (6.1) with $\epsilon = 1/(2\nu)$, $\nu = 1, 2, \cdots$. The function

$$(\rho_{1/(2\nu)} * \varphi)(x) = \int_{R^n} \rho_{1/(2\nu)}(x-t)\varphi(t)\,dt = \int_{[-l,l]} \rho_{1/(2\nu)}(x-t)\varphi(t)\,dt \tag{8.4}$$

has the property that $\partial^\sigma(\rho_{1/(2\nu)} * \varphi) = (\partial^\sigma \rho_{1/(2\nu)}) * (\varphi)$, for any n-tuple σ of nonnegative integers, belongs to $\Pi(\mathfrak{D}[-(l+1), l+1])$ and, because

$$| \partial^\sigma(\rho_{1/(2\nu)}*\varphi)(x) - \partial^\sigma\varphi(x) | = \left| \int_{R^n} \rho_{1/(2\nu)}(x-t)[\partial^\sigma\varphi(t) - \partial^\sigma\varphi(x)]\, dx \right|$$
$$\leqq \max_{t\in[x-1/(2\nu),x+1/(2\nu)]} | \partial^\sigma\varphi(t) - \partial^\sigma\varphi(x) |,$$

is seen (on applying the mean value theorem of differential calculus and the definition of $|\varphi|_p$) to satisfy

$$(8.5) \qquad | (\rho_{1/(2\nu)}*\varphi)(x) - \varphi(x) |_p \leqq \frac{n}{2\nu} | \varphi |_{p+1}$$

for every integer $p > 0$. That is, $\{\rho_{1/(2\nu)} * \varphi\}_\nu$ converges to φ in $\Pi(\mathfrak{D}[-(l+1), l+1)$ when $\nu \to \infty$ and, therefore,

$$\lim_\nu \langle F, \rho_{1/(2\nu)} * \varphi\rangle = \lim_{\nu\to\infty} (-1)^{nr}\langle G, D^r(\rho_{1/(2\nu)} * \varphi)\rangle = \langle F, \varphi\rangle.$$

Inverting the order of integration in

$$\langle G, D^r(\rho_{1/(2\nu)}*\varphi)\rangle = \int_{[-l-1,l+1]} G(x) \int_{[-l,l]} [D^r\rho_{1/(2\nu)}(x-t)]\, \varphi(t)\, dt\, dx,$$

we have $\langle F, \rho_{1/(2\nu)} * \varphi\rangle = (-1)^{nr}\langle G, D^r\rho_{1/(2\nu)} * \varphi\rangle = \langle f_\nu, \varphi\rangle$, where f_ν, defined for $t \in [-l, l]$ by

$$(8.6) \qquad f_\nu(t) = (-1)^{nr} \int_{[-l-1,l+1]} G(x)[D^r\rho_{1/(2\nu)}(x-t)]\, dx,$$

is continuous. Now, the sequence $\{f_\nu(t)\}_\nu$ has in view of (8.1) and (8.5) the property that

$$(8.7) \qquad \begin{aligned} | \langle f_\nu, \varphi\rangle | &\leqq | \langle F, \rho_{1/(2\nu)}*\varphi - \varphi\rangle | + | \langle F, \varphi\rangle | \\ &\leqq K_q \frac{n}{2\nu} | \varphi |_{q+1} + K_q | \varphi |_q \leqq K_q \left(\frac{n}{2} + 1\right) | \varphi |_{q+1}, \end{aligned}$$

which can be particularized to

$$(8.7') \qquad | \langle \alpha_j f_\nu, \varphi\rangle | \leqq K_q \left(\frac{n}{2} + 1\right) | \varphi |_{q+1}.$$

This last inequality is the same as (6.5), but with $q + 1$ in place of p. Looking over the proof of Lemma 4 it will be noticed that, while there the f_ν were $C(R^n)$ functions, only their values in $[-l, l]$ ever appeared in the computations. These computations are, therefore, applicable to the present situation. Thus, just as (6.18) follows from (6.5), so we deduce from (8.7′)

the estimate

$$\|D^{-q-2}(\alpha_j f_\nu)\|_{[-l,l]} \leqq 2K_q\left(\frac{n}{2}+1\right)C(q,l), \tag{8.8}$$

where $C(q, l)$ depends on q and l only. The arguments appearing just prior to the statements of Lemma 4 and Corollary 1 in §6 carry over verbatim (with $q + 1$ in place of p) and lead to the conclusion that, as $\nu \to \infty$, $\sum_j D^{-q-3}(\alpha_j f_\nu)$ converges uniformly to a continuous function $F_{-q}:[-l, l] \to C$ which, in view of (8.8), yields (by applying $D^{-1} = D_1^{-1}D_2^{-1} \cdots D_n^{-1}$ on both sides in (8.8) and by recalling that the integration in each variable is over an interval of length not exceeding $2l$)

$$\|F_{-q}(x)\|_{[-l,l]} \leqq 2\cdot 2^n(2l)^n\left(\frac{n}{2}+1\right)K_q\, C(q,l) \tag{8.9}$$

and proves assertion (8.2) here above. Equation (8.3) we get from

$$\langle D^{-q-3}(\alpha_j f_\nu), \varphi\rangle = (-1)^{n(q+3)}\langle \alpha_j f_\nu, D^{q+3}\varphi\rangle,$$

by summing over j and letting $\nu \to \infty$ (which is legitimate because of the uniform convergence on $[-l, l]$ of $D^{-q-3}(\alpha_j f_\nu)$. The proof of Lemma 6 is concluded.

From this we straightaway obtain Theorem 4.

THEOREM 4. *Every countable family* $\{F_\nu\}$ *of elements of* $\Delta'(\mathfrak{D}(R^n))$ *weakly bounded on the set* $\Delta(\mathfrak{D}(R^n))$ *contains a subsequence* $\{F_{\nu(i)}\}_i$ *which converges to a distribution* $L \in \mathfrak{D}'(R^n)$ (*in the sense that* $\lim_i \langle F_{\nu(i)}, \varphi\rangle = \langle L, \varphi\rangle$ *at every* $\varphi \in \mathfrak{D}(R^n)$).

One says that every weakly bounded sequence of $\Pi'(\mathfrak{D})$ functionals is relatively sequentially compact.

Proof. Obviously $\{F_\nu\}$ weakly bounded on $\Delta(\mathfrak{D}(R^n))$ implies $\{F_\nu\}$ is weakly bounded on $\Pi(\mathfrak{D}(R^n))$. By Theorem 1 each $F_\nu \in \mathfrak{D}'(R^n)$, and so $F_\nu \in \mathfrak{D}'[-(l+1), l+1]$ for every $l \geqq 1$. By Lemma 2 there exist $B_1 < \infty$ and an integer $p_2 \geqq 0$ with $|\langle F_\nu, \varphi\rangle| \leqq B_1|\varphi|_{p_2}$ when $\varphi \in \Pi(\mathfrak{D}[-(2+1), 2+1])$. Let $F_{-p_2\nu}$ be the function associated with F_ν in accordance with Lemma 6 and define

$$H_\nu(x) = \sum_j D^{-1}(\alpha_j F_{-p_2\nu});$$

we then have

$$\|H_\nu(x)\|_{[-2,2]} \leqq 2^n\cdot 4^n\, \|F_{-p_2\nu}\|_{[-2,2]},$$

$$\left\|\frac{dH_\nu(x)}{dx}\right\|_{[-2,2]} = \|F_{-p_2\nu}\|_{[-2,2]} \leqq B_1\, A(p_2, 2),$$

so that the $H_\nu(x)$ are equicontinuous on $[-2, 2]$ while

$$(8.10) \qquad \langle F_\nu, \varphi\rangle = (-1)^{n(p_2+4)} \langle H_\nu, D^{p_2+4}\varphi\rangle.$$

The Arzelà-Ascoli theorem applied to $\{H_\nu\}$ yields a subsequence $\{H_{\nu'(k)}\}_k$ converging uniformly on $[-2, 2]$ to a continuous function $E_2:[-2, 2] \to C$. It follows from (8.10) that

$$\lim_k \langle F_{\nu'(k)}, \varphi\rangle = (-1)^{n(p_2+4)}\langle E_2, D^{p_2+4}\varphi\rangle$$

when $\varphi \in \mathfrak{D}[-2, 2]$. Now, repeating this argument with $\{F_{\nu'(k)}\}$ in place of $\{F_\nu\}$ and $l = 3$ in place of $l = 2$, we find a subsequence $\{F_{\nu''(k)}\}_k$ of $\{F_{\nu'(k)}\}$ for which $\langle F_{\nu''(k)}, \varphi\rangle$ converges to $(-1)^{n(p_3+4)}\langle E_3, D^{p_3+4}\varphi\rangle$ for some $p_3 \geqq p$, E_3 continuous on $[-3, 3]$ to C, and for all $\varphi \in \mathfrak{D}[-3, 3]$. Continuing thus we obtain sequences $\{F_{\nu(l)(k)}\}_k$ with $\{F_{\nu(l+1)(k)}\}_k \subset \{F_{\nu(l)(k)}\}_k$ and continuous functions $E_l:[-l, l] \to C$ such that

$$\lim_k \langle F_{\nu(l)(k)}, \varphi\rangle = (-1)^{n(p_l+4)}\langle E_l, D^{p_l+4}\varphi\rangle$$

when $\varphi \in \mathfrak{D}[-l, l]$. Taking the diagonal sequence defined by $F_{\nu(l)} = F_{\nu(l)(l)}$ and letting L be the distribution in $\mathfrak{D}'(R^n)$ which for $\varphi \in \mathfrak{D}[-l, l]$ has the value $\langle L, \varphi\rangle = (-1)^{n(p_l+4)}\langle E_l, D^{p_l+4}\varphi\rangle$, we have the proof of the theorem.

Remark. That every sequence $\{F_\nu\} \subset \mathfrak{D}'(R^n)$ weakly bounded on $\mathfrak{D}(R^n)$ contains a convergent subsequence is well known [2], [4], [17, vol. I]. The above theorem shows that the same conclusion is reached if we merely demand that $\langle F_\nu, \varphi\rangle$ be bounded at each $\varphi \in \Pi(\mathfrak{D}(R^n))$, or even just at every $\varphi \in \Delta(\mathfrak{D}(R^n))$. The proposition remains true if we drop the countability hypothesis on $\{F_\nu\}$.

THEOREM 5. *If the sequence $\{F_\nu\}$ of elements of $\Delta'(\mathfrak{D}(R^n))$ is such that* $\lim_\nu \langle F_\nu, \varphi\rangle$ *exists at every $\varphi \in \Delta(\mathfrak{D}(R^n))$, then there exists a unique $L \in \mathfrak{D}'(R^n)$ for which* $\lim_\nu \langle F_\nu, \varphi\rangle = \langle L, \varphi\rangle$ *at every $\varphi \in \mathfrak{D}(R^n)$.*

Proof. $\{F_\nu\}$ is weakly bounded on $\Pi(\mathfrak{D}(R^n))$, so, by Theorem 4, each subsequence of $\{F_\nu\}$ contains a subsequence which converges to a $\mathfrak{D}'(R^n)$ distribution. Consider any two of them, $\{F_{\nu'(i)}\}$ and $\{F_{\nu''(i)}\}$. Then to each interval of the form $[-l, l] \subset R^n$ correspond equicontinuous functions $H_{\nu'(i)}, H_{\nu''(i)}$ with $\langle F_{\nu'(i)}, \varphi\rangle = (-1)^{nq}\langle H_{\nu'(i)}, D^q\varphi\rangle$, and $\langle F_{\nu''(i)}, \varphi\rangle = (-1)^{nq} \cdot \langle H_{\nu''(i)}, D^q\varphi\rangle$ at each $\varphi \in \Pi(\mathfrak{D}[-l, l])$, and for some integer q independent of $\nu'(i)$, $\nu''(i)$ and φ. Now, $\{H_{\nu'(i)}\}$ and $\{H_{\nu''(i)}\}$ both converge uniformly on $[-l, l]$ to the same continuous function; otherwise, an argument in all points analogous to that centered on (6.19) would exhibit a specific function $\varphi \in \Pi(\mathfrak{D}[-l, l])$ such that $\langle F_{\nu'(i)}, \varphi\rangle$ and $\langle F_{\nu''(i)}, \varphi\rangle$ do not have the same limits. This contradicts our hypotheses and concludes the proof of this proposition.

Remark. Theorem 5 is a generalization of Theorem 1 in that it extends to sequences of elements of $\Delta'(\mathfrak{D}(R^n))$ a property of sequences of functions. It also generalizes a classical result of distribution theory [2], [4], [17, vol. 1] to the effect that $\mathfrak{D}'(R^n)$ is weakly complete (i.e., if $\lim \langle F_\nu, \varphi \rangle$ exists at every $\varphi \in \mathfrak{D}(R^n)$, then there exists $L \in \mathfrak{D}'(R^n)$ for which $\lim \langle F_\nu, \varphi \rangle = \langle L, \varphi \rangle$ throughout $\mathfrak{D}(R^n)$). This theorem also nicely rounds out the interpretation of experimental measurements touched upon in §2. Indeed, an experimenter who assigns a number to a smearing function φ is rarely in a position to tell whether it is of the pure function form $\int f\varphi \, dx = \langle f, \varphi \rangle$ with $f \in V(\mathfrak{B})$ or of the more general functional form $\lim \langle f_\nu, \varphi \rangle = \langle F, \varphi \rangle$. Theorem 5 tells us that this is no serious handicap so far as the convergence of sequences is concerned: to a convergent (on $\Delta(\mathfrak{B})$) sequence $\{F_\nu\}$ of functionals there always corresponds a sequence of functions $\{f_\nu\}$ such that $\lim \langle f_\nu, \varphi \rangle = \lim \langle F_\nu, \varphi \rangle$ not only at every $\varphi \in \Delta(\mathfrak{B})$ but at every $\varphi \in \mathfrak{B}$ as well.

What has been demonstrated here for sequences of elements of $\Delta'(\mathfrak{D}(R^n))$ can also be shown for sequences of elements in $\Delta'(\mathcal{S})$ or $\Delta'(S_{\alpha,A})$. The proofs are sufficiently similar to those just given (with Lemma 5 replacing Lemma 4) to spare us the need to spell them out here. It suffices to state the results. (In each one of them $\mathfrak{B}_F$ stands for $\mathcal{S}(R^n)$ or $S_{\alpha,A}(R^n)$, at will.)

THEOREM 6. *Every sequence $\{F_\nu\} \subset \Delta'(\mathfrak{B}_F)$ weakly bounded on $\Delta(\mathfrak{B}_F)$ contains a subsequence $\{F_{\nu(i)}\}_i$ such that $\lim_i \langle F_{\nu(i)}, \varphi \rangle = \langle L, \varphi \rangle$ for some $L \in \mathfrak{B}_F'$ and all $\varphi \in \mathfrak{B}_F$.*

THEOREM 7. *If the sequence $\{F_\nu\} \subset \Delta'(\mathfrak{B}_F)$ converges at every $\varphi \in \Delta(\mathfrak{B}_F)$ (i.e., $\lim \langle F_\nu, \varphi \rangle$ exists $\forall\, \varphi \in \Delta(\mathfrak{B}_F)$), there exists a unique $L \in \mathfrak{B}_F'$ such that $\lim \langle F_\nu, \varphi \rangle = \langle L, \varphi \rangle$ at every $\varphi \in \mathfrak{B}_F$.*

9. Concluding observations. Starting from some simple physical considerations we were led to the view that certain functionals on sets of probability densities are natural constructs for the theoretical description and correlation of experimental data. These functionals were shown to be distributions, after which it was seen that these distributions are entirely determined (both as to identification and convergence properties) by their behavior on the subset of test functions consisting of probability densities.

Applications of these results, e.g., to defining the concept of "value at a point" for a distribution and relating it to earlier studies of this notion [11] and to the physical aspects of "ill-posed" problems in partial differential equations, will be examined elsewhere.

REFERENCES

[1] P. W. BRIDGMAN, *The Nature of Physical Theory*, Princeton University Press, Princeton, 1936.

[2] A. Friedman, *Generalized Functions and Partial Differential Equations*, Prentice-Hall, Englewood Cliffs, New Jersey, 1963.

[3] I. M. Gelfand and G. E. Šilov, *Quelques applications de la théorie des fonctions généralisées*, J. Math. Pures Appl. (9), 35(1956), pp. 383–413.

[4] ——, *Les distributions. Espaces fondamentaux*, vol. II, Dunod, Paris, 1964.

[5] P. K. Ghosh, Appendix to *Direct Analysis of Diffraction by Matter*, by R. Hosemann and S. N. Bagchi, North Holland, Amsterdam, 1962.

[6] I. Halperin, *Introduction to the Theory of Distributions*, University of Toronto Press, Toronto, 1952.

[7] R. L. Ingraham, *Stochastic Lorentz observers and the divergencies in quantum field theory*, Nuovo Cimento, 24 (1962), pp. 1117–1146. (Corrections and emendations: Ibid., 27 (1963), pp. 303–305.)

[8] ——, *Stochastic space-time*, Ibid., 31 (1964), pp. 512–529.

[9] M. J. Lighthill, *Introduction to Fourier Analysis and Generalized Functions*, Cambridge University Press, New York, 1958.

[10] T. P. G. Liverman, *Generalized Functions and Direct Operational Methods*, vol. I, Prentice-Hall, Englewood Cliffs, New Jersey, 1964.

[11] S. Łojasiewicz, *Sur la valeur et la limite d'une distribution en un point*, Studia Math., 16(1957), pp. 1–36.

[12] S. Mandelbrojt, *Analytic functions and classes of infinitely differentiable functions*, Rice Institute Pamphlet 29, no. 1, 1942, 142 pp.

[13] G. Marinescu, *Espaces vectoriels pseudo-topologiques et théorie des distributions*, VEB, Berlin, 1963.

[14] J. G. Mikusiński, *Sur la méthode de généralisation de M. Laurent Schwartz et sur la convergence faible*, Fund. Math., 35(1948), pp. 235–239.

[15] J. R. Ravetz, *Distributions defined as limits*, Proc. Cambridge Philos. Soc., 53(1957), pp. 76–92.

[16] F. Riesz and B. Sz. Nagy, *Leçons d'analyse fonctionnelle*, 3rd ed., Académie des Sciences de Hongrie, Akadémiai Kiadó, Budapest, 1955.

[17] L. Schwartz, *Théorie des Distributions*, vols. I, II, Hermann, Paris, 1950–1951.

[18] G. Temple, *The theory of generalised functions*, Proc. Roy. Soc. London Ser. A, 228 (1955), pp. 175–190.

[19] A. H. Zemanian, *Distribution Theory and Transform Analysis*, McGraw-Hill, New York, 1965.

[20] ——, *The convolution transformation of certain generalized functions and its inversion*, Bull. Amer. Math. Soc., 72 (1966), pp. 725–728.

[21] ——, *A generalized Weierstrass transformation*, this Journal, 15(1967), pp. 1088–1105.

THE CAUCHY INTEGRAL OF TEMPERED DISTRIBUTIONS AND SOME THEOREMS ON ANALYTIC CONTINUATION*

E. J. BELTRAMI† AND M. R. WOHLERS‡

1. Introduction. A continuation formula for holomorphic functions that map a half-plane into itself and, in particular, for positive real functions was recently established [1] (see Corollary 3.1 in §3 for a precise statement of this result). Such functions arise in a natural way in studying linear passive systems and, in fact, Dolph assigned an important role to this result in his investigation of maximally dissipative operators on a Hilbert space since the resolvents of such operators are, in a sense, positive real [2].

In this paper, we extend the continuation formula to a larger class of holomorphic functions, namely, to the class H^+ of functions that are holomorphic in $\text{Re}\, p > 0$ and of polynomial growth uniformly in every half-plane $\text{Re}\, p \geqq \sigma > 0$. Such functions (which contain the positive real class) come about in studying causal, but not necessarily passive, linear systems and characterize a class of Laplace transforms of Schwartz distributions (cf. the remarks in §3).

Our approach to the continuation problem will be in terms of the Cauchy integral of tempered distributions that are defined in §2. A number of results concerning such Cauchy integrals are summarized in Theorem 2.1 since they are of independent interest. In particular, we display the explicit boundary behavior of such Cauchy integrals and their derivatives. The main argument used here requires that all calculations with tempered distributions be reduced to a more restricted class of objects by means of a suitable division (Lemma 2.1).

Crucial to the proof of the continuation theorem in [1] is a representation for positive real functions due to Herglotz and Cauer. Those H^+ functions having boundary values in the Schwartz topology of tempered distributions can be represented by a Cauchy integral of the real part of the boundary value, a result that enjoys the same role as the Cauer formula does in the positive real case. It is this fact that allows us to establish Theorem 3.1, the main result of §3.

A final section, §4, is concerned with showing how the notion of Cauchy

* Received by the editors October 26, 1966, and in revised form February 8, 1967. Contributed at the Symposium on "The Applications of Generalized Functions" sponsored by the Air Force Office of Scientific Research at the 1966 Fall Meeting of Society for Industrial and Applied Mathematics held at the State University of New York at Stony Brook, September 12–14, 1966.

† State University of New York at Stony Brook, Stony Brook, Long Island, New York 11790. This author's research was supported by the Air Force Office of Scientific Research under Contract AF-AFOSR-1154-66.

‡ Grumman Aircraft Engineering Corporation, Bethpage, New York.

integral for distributions can be utilized to factor holomorphic functions in a strip (Theorem 4.1). This factorization is achieved by a suitable continuation into overlapping half-plane functions and has application to the Wiener-Hopf problem.

2. The Cauchy integral of tempered distributions. In what follows, we will use the notation and terminology of distribution theory as given by L. Schwartz [3]. Some of the exceptions are explicitly noted below.

Let C_0^∞ denote the collection of all complex-valued infinitely differentiable functions having compact support on the real axis. The closure of C_0^∞ in the norm $(\sum_{j\leq m}\int_{-\infty}^{\infty} | D^j\phi |^2)^{1/2}$ is a reflexive B-space, the Sobolev space $W^{m,2} \subset L_2$ (m any nonnegative integer). Then u belongs to $W^{m,2}$ if and only if the weak or distributional derivatives $D^j u$ belong to $L_2(-\infty, \infty)$ or simply L_2 for all $j \leqq m$. The strong or norm dual of $W^{m,2}$ is denoted by $W^{-m,2}$, and $v \in W^{-m,2}$ if and only if $v = \sum_{j\leqq m} D^j v_j$, where $v_j \in L_2$. (For details on Sobolev spaces and their duals, see [4, Chap. 1].) Now let v_t be a tempered distribution (Schwartz class S') in the real variable t. Then the distributional Fourier transform of v_t is denoted by $\mathfrak{F}v_t \equiv u_\omega$, and we have the inversion $\mathfrak{F}^{-1}u_\omega = v_t$. The Laplace transform of any $v_t \in S' \cap \mathfrak{D}_+'$ (tempered distributions with support in the positive t-axis) is defined by $\mathfrak{F}(v_t e^{-\sigma t}) \equiv U(p)$ for $\sigma > 0$, with $p = \sigma + i\omega$. Finally, we have, topologically, $L_2 \subset W^{-m,2} \subset S'$.

We begin by proving a lemma.

LEMMA 2.1. *Let $u_\omega \in S'$. Then there exist integers $m, l \geqq 0$ such that $(i\omega)^l f_\omega = u_\omega$ has a solution $f_\omega \in W^{-m,2}$. Moreover, the division is such that whenever $\mathfrak{F}^{-1}u_\omega \in S' \cap \mathfrak{D}_+'$ then f_ω is uniquely determined if we impose the condition that $\mathfrak{F}^{-1}f_\omega \in S' \cap \mathfrak{D}_+'$ as well.*

Proof. We can write, for any $u_\omega \in S'$, $\mathfrak{F}^{-1}u_\omega = v_t = D^l g_t$ for some weak lth order derivative where g_t is a continuous function of polynomial growth [3, vol. 2, p. 95]. But then $u_\omega = (i\omega)^l f_\omega$, where $f_\omega = \mathfrak{F}g_t \in W^{-m,2}$ for some $m \geqq 0$ (see [3, vol. 2, p. 112]). Now, f_ω is not uniquely determined since $(i\omega)^l f_\omega = 0$ implies that f_ω is a finite sum of δ and its derivatives. However, when $v_t \in S' \cap \mathfrak{D}_+'$, then g_t can also be chosen in a unique way to have its support in the same positive t-axis (see [4, Theorem 1.37]); so f_ω will be a unique solution to $(i\omega)^l f_\omega = u_\omega$ if we demand that $\mathfrak{F}^{-1}f_\omega \in \mathfrak{D}_+'$.

In what follows we assume, without further stating it, that f_ω is uniquely determined in this manner whenever $\mathfrak{F}^{-1}u_\omega \in S' \cap \mathfrak{D}_+'$.

Now, let $f_\omega \in W^{-m,2}$. Then, since $1/(p - i\zeta) \in W^{m,2}$ as a function of ζ for each fixed complex $p = \sigma + i\omega$ with $\sigma \neq 0$, the expression $\langle f_\zeta, 1/(p - i\zeta)\rangle$ is well defined. Moreover, a simple argument shows that $\langle f_\zeta, 1/(p - i\zeta)\rangle$ is holomorphic in the half-planes $\operatorname{Re} p \neq 0$ (see, for example, [4, p. 59]). Now, if $u_\omega \in S'$ and if f_ω is a division of u_ω (Lemma 2.1), then we define the Cauchy integral of u_ω to be

$$C(u, p) = \frac{p^l}{2\pi}\left\langle f_\zeta, \frac{1}{p - i\zeta}\right\rangle. \tag{2.1}$$

From what we said above, $C(u, p)$ exists and is holomorphic for $\operatorname{Re} p \neq 0$. Whenever $u_\omega \in W^{-m,2}$ for any $m \geqq 0$, there is no need to divide it by $(i\omega)^l$ and so $l = 0$ for all such distributions; in this case $C(u, p)$ is simply

$$\frac{1}{2\pi}\left\langle u_\zeta, \frac{1}{p - i\zeta}\right\rangle,$$

and the Cauchy integral of any $u_\omega \in S'$ can then be written as $p^l C(f, p)$, where f_ω is the distribution $u_\omega/(i\omega)^l$.

If $u_\omega \in W^{-m,2}$ itself, in which case $l = 0$, or if $\mathfrak{F}^{-1}u_\omega \in S' \cap \mathfrak{D}_+'$, then, as we saw, f_ω and hence $C(u, p)$ will be uniquely determined. Otherwise, there is a plenitude of Cauchy integrals for any given u_ω, all of which differ at most by a polynomial in p. In what follows, the division of any given $u_\omega \in S'$ is arbitrary, but fixed, except in the case where $\mathfrak{F}^{-1}u_\omega \in S' \cap \mathfrak{D}_+'$ (see remark after Lemma 2.1).

It is clear that $C(u, p)$ coincides with the usual half-plane Cauchy representation whenever $u_\omega \in L_2$. Actually, the analogy with L_2 lies much deeper as we can see from the next theorem, which extends a number of L_2 results (see the remarks after the proof). The theorem itself summarizes several of the important properties of $C(u, p)$ for $u_\omega \in S'$.

THEOREM 2.1. *Let $u_\omega \in S'$ and let $f_\omega \in W^{-m,2}$ be defined as in Lemma 2.1. Then, as $\sigma \to 0$, we obtain the extended Plemelj relations*

$$C(u,p) \to u_\omega^+ = \frac{u_\omega}{2} + \frac{(i\omega)^l}{2\pi} f_\omega * \operatorname{pf}\frac{1}{\omega}, \qquad \sigma > 0,$$

$$-C(u,p) \to u_\omega^- = \frac{u_\omega}{2} - \frac{(i\omega)^l}{2\pi} f_\omega * \operatorname{pf}\frac{1}{\omega}, \qquad \sigma < 0,$$

where the limits are taken in the S' topology and pf *denotes Hadamard finite part. In particular, we obtain the Hilbert decomposition $u_\omega = u_\omega^+ + u_\omega^-$, where u_ω^+ and u_ω^- are Fourier transforms of distributions having their support in the positive and negative half-axis, respectively. Moreover, for any $j \geqq l$,*

$$\begin{aligned} \frac{d^j}{dp^j} C(u,p) &= \frac{(-1)^j j!}{2\pi}\left\langle u_\zeta, \frac{1}{(p - i\zeta)^{j+1}}\right\rangle \\ &= \frac{(i)^j}{2\pi}\left\langle u_\zeta, \frac{d^j}{d\zeta^j}\frac{1}{p - i\zeta}\right\rangle. \end{aligned} \tag{2.2}$$

Now let H^+ *denote the collection of functions holomorphic in the half-plane* $\operatorname{Re} p > 0$ *that are uniformly bounded by a polynomial on every half-plane* $\operatorname{Re} p \geqq \sigma > 0$ *(the bound depends on* σ*). Then* u_ω *is the boundary value taken in the* S' *topology as* $\sigma \to 0$ *of an* H^+ *function* $f(p)$ *if and only if*

(i) $\mathfrak{F}^{-1}u_\omega = v_t \in S' \cap \mathfrak{D}_+'$,

in which case $f(p) = U(p)$, *the Laplace transform of* v_t. *Alternate necessary and sufficient conditions for* (i) *to hold are then*

(ii) $C(u, p) = 0, \quad \sigma < 0,$

(iii) $C(u, p) = U(p), \quad \sigma > 0,$

(iv) $D^j u_\omega = \dfrac{1}{\pi i} u_\omega * D^j \operatorname{pf} \dfrac{1}{\omega} = \dfrac{(-1)^j j!}{\pi i} u_\omega * \operatorname{pf} \dfrac{1}{\omega^{j+1}}$ *for all* $j \geqq m$,

(v) $u_\omega = u_\omega^+$.

Finally, if any of (i) *through* (v) *hold, then*

$$(2.3) \quad C(u, p) = \frac{p^l}{\pi}\left\langle \operatorname{Re} f_\zeta, \frac{1}{p - i\zeta}\right\rangle = \frac{ip^l}{\pi}\left\langle \operatorname{Im} f_\zeta, \frac{1}{p - i\zeta}\right\rangle.$$

Proof. By virtue of Lemma 2.1, it suffices to establish the theorem for $u_\omega \in W^{-m,2}$ (see [4, Chap. 3]). To illustrate the method of reasoning, we sketch through the argument for (2.2):

$$\frac{d^j}{dp^j} C(u, p) = \frac{1}{2\pi} \frac{d^j}{dp^j}\left\langle f_\zeta, \frac{p^l}{p - i\zeta}\right\rangle$$

and, by a continuity argument, this expression equals

$$\frac{1}{2\pi}\left\langle f_\zeta, \frac{d^j}{dp^j} \frac{p^l}{p - i\zeta}\right\rangle.$$

But

$$\frac{d^j}{dp^j}\left(\frac{p^l}{p - i\zeta}\right) = (i\zeta)^l \frac{d^j}{dp^j} \frac{1}{p - i\zeta}$$

for all $j \geqq l$, and so

$$\begin{aligned} \frac{d^j}{dp^j} C(u, p) &= \frac{(-1)^j j!}{2\pi}\left\langle f_\zeta, \frac{(i\zeta)^l}{(p - i\zeta)^{j+1}}\right\rangle \\ &= \frac{(-1)^j j!}{2\pi}\left\langle u_\zeta, \frac{1}{(p - i\zeta)^{j+1}}\right\rangle \\ &= \frac{(i)^j}{2\pi}\left\langle u_\zeta, \frac{d^j}{d\zeta^j} \frac{1}{p - i\zeta}\right\rangle. \end{aligned}$$

The first two identities in (2.2) are well known for $u_\omega \in L_2$. Finally, let us emphasize that if $u_\omega \in W^{-m,2}$, then the statements of the theorem hold with $l = 0$ and hence with $f_\omega \equiv u_\omega$. Moreover, the Cauchy integral is uniquely defined in this case.

Remarks. Condition (i) extends the L_2 one-sided Paley-Wiener theorem since the Hardy class H^2 belongs to the collection of H^+ functions which have S' boundary values; moreover, as we know, $L_2(0, \infty) \subset S' \cap \mathfrak{D}_+'$. The Paley-Wiener theorem expresses the causality of certain linear systems. Thus, it is not surprising that the extended condition (i) is also a statement of causality for a larger class of systems (see [5] for a discussion). The remaining alternate conditions (ii) through (v) are well known in the L_2 case, and, in particular, (iv) is equivalent to the fact that the real and imaginary parts of $D^j u_\omega$ satisfy reciprocal relations (dispersion relations in physics; see [5] for a discussion) for all $j \geqq l$. In fact, when $u_\omega \in W^{-m,2}$, we can let $j = 0$ and (iv) then tells us that the conjugate Hilbert transforms

$$\operatorname{Re} u_\omega = \frac{1}{\pi} \operatorname{Im} u_\omega * \operatorname{pf} \frac{1}{\omega},$$

$$\operatorname{Im} u_\omega = -\frac{1}{\pi} \operatorname{Re} u_\omega * \operatorname{pf} \frac{1}{\omega}$$

hold, in complete analogy with the L_2 case. Condition (iv) can also be viewed as the Cauchy derivative formula taken to the boundary. We see this immediately whenever $u_\omega \in W^{-m,2}$; for then from (2.2) and since $D^j u \in W^{-(m+j),2}$ we have

$$\frac{d^j}{dp^j} C(u, p) = \frac{(-i)^j}{2\pi} C(D^j u, q)$$

$$\to (-i)^j D^j u_\omega = \frac{(-i)^l}{\pi i} u_\omega * D^j \operatorname{pf} \frac{1}{\omega}$$

as $\sigma \to 0$. But

$$D^j \operatorname{pf} \frac{1}{\omega} = \operatorname{pf} D^j \frac{1}{\omega} = (-1)^j j! \operatorname{pf} \frac{1}{\omega^{j+1}}$$

and so

$$D^j u_\omega = \frac{(-i)^j j!}{\pi i} u_\omega * \operatorname{pf} \frac{1}{\omega^{j+1}}.$$

In the case where $v_t = D^l g_t$ with the support of g_t in the half-axis $t \geqq 0$, we have $C(u, p) = p^l C(f, p) = p^l V(p) = U(p)$, where $V(p)$ is the Laplace transform of g_t. Assertion (iii) follows from this fact once we know that $V(p) = C(f,p)$ for $\sigma > 0$; the last statement is established in [4], as already noted.

For Re $p < 0$, the Cauchy integral $C(u, p)$ is equal to $C(u, -\bar{p})$ when Re $p > 0$. This allows us to define the Poisson integral $C(u, p) - C(u, -\bar{p})$, $\sigma > 0$. Then, from the Plemelj relations of Theorem 2.1, the Poisson integral tends to u_ω in the S' topology as $\sigma \to 0$. Note that (ii) is the same as saying that the Cauchy and Poisson integrals are identical for $\sigma > 0$.

As we will see in §3, (2.3) is analogous to the Herglotz-Cauer representation for certain holomorphic functions with positive real part in a half-plane since, in this case, the Cauchy integral is in terms of a suitable measure which is the real part of the boundary value. This fact will be basic to us in proving the results of the next section. If $u_\omega \in W^{-m,2}$, then from (2.3) we obtain $C(u, p) = 2C(\text{Re } u, p) = 2iC(\text{Im } u, p)$.

The notion of Cauchy integral has also been studied by H. Bremermann and L. Durand [6] for a certain class of distributions. The results are summarized in Bremermann's recent book [7], as is a discussion of the Fourier transform of tempered distributions from a different point of view. Indeed, some of the results to be found in [7] overlap the topics covered in the section above although the treatment is different. The present authors introduced the Cauchy integral for distributions in $W^{-m,2}$ in [8] and later in [9] for S' objects; parts of Theorem 2.1 were proven in [9] although some of the results had already been established in the extended settings of tubular domains in C^n by L. Garding (unpublished, see [10]), N. Vladimirov [11], and Bremermann [7], as we were to learn later. In fact, the collection of all H^+ functions having S' boundary values has been thoroughly studied in C^n by Vladimirov, who designated them by T^+. A thorough discussion is available in [4]. The classical version of Theorem 2.1 can be found in the books by Muskhelishvili [12] and, particularly, Titchmarsch [13, Chap. 5].

3. A continuation theorem. The principal result of this section is given by the next theorem.

THEOREM 3.1. *A necessary and sufficient condition that the H^+ function $U(p)$, having S' boundary value U_ω as $\sigma \to 0$, be analytically continuable across an open set Ω of the ω-axis, is that* Re U_ω *be analytic on Ω. The extended holomorphic function is then given by*

$$U(p) = -\overline{U(-\bar{p})} + 2g(p) \tag{3.1}$$

for Re $p < 0$, *where $g(p)$ is the analytic extension of* Re u_ω *as given on Ω.*

Proof. It suffices to consider the case where Ω is an interval. Suppose that $U(p)$ is continuable. Then $U(p)$ is defined and holomorphic in a domain D_1 which includes Ω; similarly, the reflected $\overline{U(-\bar{p})}$ is holomorphic in the reflected domain D_2 which also includes Ω. Therefore, $g(p) = [U(p) + \overline{U(-\bar{p})}]/2$ is holomorphic on $D = D_1 \cap D_2$. But $g(i\omega) = \operatorname{Re} U(i\omega)$, which proves necessity (note that $g(p)$ is not equal to $\operatorname{Re} U(p)$ except on Ω).

To establish the converse, we first note that

$$D^l C(u, p) = \text{const.} \left\langle u_\zeta, \frac{1}{(p - i\zeta)^{l+1}} \right\rangle$$

for some $l \geqq 0$ and $\operatorname{Re} p \neq 0$ as we saw above (Theorem 2.1), and so, using the arguments used to establish (2.3), we obtain

$$D^l C(u, p) = \lambda \left\langle \operatorname{Re} u_\zeta, \frac{1}{(p - i\zeta)^{l+1}} \right\rangle \tag{3.2}$$

for some constant λ. Now let $v_\omega = \operatorname{Re} u_\omega - \chi \operatorname{Re} u_\omega \in S'$, where χ is the characteristic function of Ω' for any closed interval $\Omega' \subset \Omega$. Then

$$D^l C(u, p) - \lambda \left\langle \chi_\zeta \operatorname{Re} u_\zeta, \frac{1}{(p - i\zeta)^{l+1}} \right\rangle = \lambda \left\langle v_\zeta, \frac{1}{(p - i\zeta)^{l+1}} \right\rangle$$

is certainly continuable across the interior of Ω' since $v_\omega = 0$ on Ω'. It suffices to show now that

$$\int_{\Omega'} \frac{\operatorname{Re} u(i\zeta)}{(p - i\zeta)^{l+1}}\, d\zeta \equiv \left\langle \chi_\zeta \operatorname{Re} u_\zeta, \frac{1}{(p - i\zeta)^{l+1}} \right\rangle \tag{3.3}$$

is also continuable. To this end, note that by hypothesis $U(i\omega)$ can be extended to be analytic on D, and so we can deform the contour defined by Ω' into some path in the left half-plane but still remain within D. Then (3.3) remains unchanged, but the continuation is now immediate. Hence, $D^l U(p)$ can be extended and so, of course, can $U(p)$. Since $\Omega' \subset \Omega$ is arbitrary, we are through. To obtain formula (3.1), consider $H(p) = U(p) - g(p)$, which is holomorphic on D and purely imaginary on Ω. Then $-\overline{H(-\bar{p})}$ is also analytic on D and has the same value on Ω. The Schwartz reflection argument shows that for $p \in D$, we have

$$U(p) = -\overline{U(-\bar{p})} + g(p) + \overline{g(-\bar{p})}.$$

But $g(p)$ is itself obtained by reflection and so $\overline{g(-\bar{p})} = g(p)$. Therefore, (3.1) holds, and it allows us to compute $U(p)$ in the left-hand plane from its values in the right-hand plane and from $g(p)$ which is the known continuation of $\operatorname{Re} u_\omega$ on Ω.

We want now to specialize Theorem 3.1 to the case where $U(p)$ is positive real.

DEFINITION 3.1. $U(p)$ is *positive real* if, in $\operatorname{Re} p > 0$, $U(p)$ is holomorphic and $\operatorname{Re} U(p) \geqq 0$; moreover, we require $U(p)$ to be real for p real.

LEMMA 3.1. *Let $U(p)$ be positive real. Then $U(p)$ belongs to H^+ and has an S' boundary value u_ω as $\sigma \to 0$.*

Proof. Analytic functions in a disk with positive real part can be represented by a Cauchy type integral in terms of a measure on the boundary. This result of Herglotz was extended by Cauer [14] to positive real functions and reads:

$$U(p) = Ap + p \int_{-\infty}^{\infty} \frac{(1+\zeta^2)}{p^2+\zeta^2}\, d\mu(\zeta), \tag{3.4}$$

where $A \geqq 0$ and μ is a real odd nondecreasing bounded measure. The fact that (3.4) belongs to H^+ and has an S' boundary value involves a classical convergence argument and may be found in [15].

COROLLARY 3.1 (Greenstein [1]). *Let $U(p)$ be positive real. Then $U(p)$ can be continued across Ω if and only if the measure $\mu(\omega)$ of (3.4) is analytic on Ω. The continuation is then given by*

$$U(p) = \overline{U(-\bar{p})} + 2\pi(1-p^2)D\mu(p). \tag{3.5}$$

Proof. In [15], the explicit boundary value u_ω of $U(p)$ is computed. In particular, we find there that $\langle \operatorname{Re} u_\omega, \phi(\omega)\rangle = \pi \int_{-\infty}^{\infty} \phi(\omega)(1+\omega^2)\, d\mu(\omega)$ for all $\phi \in S$, and so $\operatorname{Re} u_\omega$ determines the real-valued measure ψ defined by

$$\psi(\omega) = \pi \int_0^{\omega} (1+\zeta^2)\, d\mu(\zeta). \tag{3.6}$$

By Theorem 3.1, the continuation is possible if and only if ψ is analytic on Ω and hence if and only if μ is analytic. Moreover, $\operatorname{Re} u_\omega = \pi(1+\omega^2)D\mu(\omega)$. The continuation is then given by (3.1) as $U(p) = \overline{U(-\bar{p})} + 2D\psi(p)$, which is (3.5).

Remarks. Positive real functions arise as the immittance or Laplace transforms of the Green's kernel of certain passive operators. In fact, they completely characterize linear, time-invariant systems, such as passive networks, which can be described by such operators (see [16, Chap. 10] for a discussion). Moreover, Dolph showed that a class of maximally dissipative operators A on a Hilbert space H_0 also describe a class of passive systems, and the resolvent $R_p(A)$ is such that $(R_p(A)u, u)$ is positive real for all $u \in H_0$ (see [2]). The significance of the continuation formula is that it allows a penetration of positive real functions into the left-hand plane where the complex singularities of the operator are to be found. Now, linear passive systems are also causal; and so, the Laplace transform of the resulting Green's kernel, which belongs to $S' \cap \mathfrak{D}_+'$, is holomorphic in $\operatorname{Re} p > 0$ (Theorem 2.1). If the system is causal but not necessarily passive, then under suitable conditions the kernel still belongs to $S' \cap \mathfrak{D}_+'$, and so its Laplace transform $U(p)$ belongs to H^+ although it no longer need be positive-real. Theorem 3.1 extends the continuation result to this larger class of systems.

In the proof of Corollary 3.1, we identify the distribution Re u_ω with the measure ψ of (3.5), whereas Greenstein uses the less explicit inversion formula

$$\psi(\omega_2) - \psi(\omega_1) = \lim_{\sigma\to 0} \int_{\omega_1}^{\omega_2} \operatorname{Re} U(\sigma + i\omega)\, d\omega.$$

It is worth noting that the continuation could also have been given in terms of Im u_ω since (3.2) is still valid in this case although with a different constant λ (i.e., (3.2) with Re u_ω replaced by Im u_ω). In the positive real case, no such continuation result would be possible classically, for, although Re u_ω is a measure, the distribution Im u_ω cannot itself be identified with any measure. These remarks allow for an essential improvement of Corollary 3.1.

It appears likely that Theorem 3.1 can be further extended to the cases where (i) $U(p)$ is the Laplace transform of a tempered distribution with support in the forward cone $\Gamma^+ \subset R^n$ (in which case $U(p)$ is holomorphic in the tube $\Gamma^+ + iR^n$ in C^n, [10]), and (ii) $U(p)$ is a positive real matrix (corresponding, for example, to n-port networks). Detailed results are not yet available.

Finally we note there is a stronger version of Theorem 3.1 which reads as follows.

THEOREM 3.2. *Let $U(p)$ be holomorphic in a domain D of the right half-plane, and suppose that the open set Ω of the ω-axis is on the boundary of D. A necessary and sufficient condition that $U(p)$ be continuable through Ω is that $U(p)$ have a boundary value u_ω in the $\mathfrak{D}'$ Schwartz topology and that* Re u_ω *by analytic on Ω.*

Proof. Necessity is as in Theorem 3.1. Conversely, let $g(p)$ be the extension of Re u_ω, and form $h(p) = -\overline{U(-\bar{p})} + 2g(p)$ for Re $p < 0$; note that $h(p)$ tends to u_ω as $\sigma \to 0$ from the left. The same boundary value is attained by $U(p)$ from the right. Hence, by the "edge of the wedge" theorem (cf. [10]), $h(p)$ is the continuation of $U(p)$.

The "edge of the wedge" theorem is a nontrivial result also established by arguments involving Cauchy integrals. The direct computational proof given in Theorem 3.1 allowed us to avoid this deeper result.

4. Remarks on the Wiener-Hopf factorization. In this section, we will comment briefly on another application of the Cauchy integral of tempered distributions. Consider the problem of solving the Wiener-Hopf equation,

$$\int_0^\infty \kappa(t-\tau)u(\tau)\, d\tau = f(t), \qquad t \geqq 0. \tag{4.1}$$

If the Laplace transform $K(p)$ of κ is holomorphic in some strip defined by the interval $\Gamma: \sigma_1 \leqq \operatorname{Re} p \leqq \sigma_2$ (which will be true whenever $\kappa_t e^{-\sigma t} \in S'$ for $\sigma \in \Gamma$), then an approach to solving (4.1) leads to factoring $K(p)$ into $K^+(p)K^-(p)$, where K^+, K^- are holomorphic in the respective half-planes defined by $\operatorname{Re} p > \sigma_1$ and $\operatorname{Re} p < \sigma_2$ (see, for example, the book by Noble [17]). One way of doing this, which extends the scope of the classical procedure is to apply the next theorem to $\log K(p)$ to obtain an analytic continuation into overlapping half-planes given by $\log K^+(p) + \log K^-(p)$. Then the required factorization is found by taking $\exp(\log K(p))$.

THEOREM 4.1. *Let $f(p)$ be holomorphic in the strip defined by* $\Gamma: \sigma_1 \leqq \operatorname{Re} p \leqq \sigma_2$. *Suppose $f(p)$ is of polynomial growth uniformly in every interior strip and that $f(\sigma + i\omega) \to f_\omega^+$ in the S' topology as $\sigma \to \sigma_1$, while $f(\sigma + i\omega) \to f_\omega^-$ when $\sigma \to \sigma_2$. All these conditions will be satisfied if $f(p)$ is the Laplace transform of some v_t for which $v_t e^{-\sigma t} \in S'$ when $\sigma \in \Gamma$; then $f_\omega^+ = \mathfrak{F}(v_t e^{-\sigma_1 t})$ and $f_\omega^- = \mathfrak{F}(v_t e^{-\sigma_2 t})$.* (See [4, Chap. 2] for details.) *Then*

$$f(p) = C(f_\omega^+, p) - C(f_\omega^-, p) = f^+(p) + f^-(p),$$

where $f^+(p)$ and $f^-(p)$ are holomorphic in the respective half-planes $\operatorname{Re} p > \sigma_1$ *and* $\operatorname{Re} p < \sigma_2$.

Proof. Since $f_\omega^+, f_\omega^- \in S'$, we form the Cauchy integrals $f^+(p) = C(f^+, p)$ and $f^-(p) = -C(f^-, p)$. That these Cauchy expressions form the desired decomposition is proven by the arguments given in [4, Chap. 2] where a similar decomposition is given in terms of Laplace transforms which are holomorphic in the required overlapping half-planes. Since Theorem 2.1 identifies the Cauchy and Laplace representations in such cases (with the half-planes shifted to $\sigma = 0$ there), we are through.

Remarks. The classical procedure requires, for example, that $\log K(p)$ be $O(|\omega|^{-\alpha})$, $\alpha > 0$, as $|\omega| \to \infty$ uniformly in any interior strip so that the additive decomposition of $\log K(p)$ be well defined by the Cauchy integral theorem (i.e., a rectangular contour is formed within the strip with two sides parallel to the edges; the Cauchy integral along the end segments vanishes as we elongate the figure to infinity, and we are left with a sum of two Cauchy integrals along the remaining contours; these integrals define the required half-plane functions). In our approach, we admit a larger class of functions which are $O(|\omega|^l)$ for any $l \geqq 0$ as $|\omega| \to \infty$, uniformly in any interior strip. Moreover, the classical representation is refined in our case by allowing the Cauchy integrals to be taken along the edge of the strip with respect to the distributional boundary values. This is the content of Theorem 4.1. Then, if Γ degenerates into a line, we obtain the additive Hilbert decomposition of Theorem 2.1. What this suggests is that, for any degenerate Wiener-Hopf or Riemann-Hilbert problem (i.e., when the strip is degenerate), we might tighten the conditions on the kernel κ sufficiently to allow for some nontrivial strip defined by Γ. We then proceed as indicated above to form an appropriate factorization within the strip. If we now let Γ shrink again, then, under suitable conditions, this should allow us to define, in the limit, the logarithm and hence the factorization of a certain class of distributions. This problem remains unresolved except in special cases [12]. Another open question is to seek conditions on $K(p)$ or, equivalently, on the kernel κ so as to guarantee that $\log K(p)$ satisfies the conditions of Theorem 4.1.

REFERENCES

[1] D. S. Greenstein, *On the analytic continuation of functions which map the upper half-plane into itself*, J. Math. Anal. Appl., 1 (1960), pp. 355–362.

[2] C. L. Dolph, *Positive real resolvents and linear passive Hilbert systems*, Ann. Acad. Sci. Fenn. Ser. A I, 336/9 (1963), pp. 331–339.

[3] L. Schwartz, *Théorie des distributions*, vols. I, II, Hermann, Paris, 1957–1959.

[4] E. J. Beltrami and M. R. Wohlers, *Distributions and the Boundary Values of Analytic Functions*, Academic Press, New York, 1966.

[5] E. J. Beltrami, *Linear dissipative systems, nonnegative definite distributional kernels, and the boundary values of bounded-real and positive-real matrices*, J. Math. Anal. Appl., to appear.

[6] H. J. Bremermann and L. Durand, *On analytic continuation, multiplication, and Fourier transformations of Schwartz distributions*, J. Mathematical Phys., 2 (1961), pp. 240–258.

[7] H. J. Bremermann, *Distributions, Complex Variables, and Fourier Transforms*, Addison-Wesley, Reading, Massachusetts, 1965.

[8] E. J. Beltrami and M. R. Wohlers, *Distributional boundary value theorems and Hilbert transforms*, Arch. Rational Mech. Anal., 18 (1965), pp. 304–309.

[9] ———, *Distributional boundary values of functions holomorphic in a half-plane*, J. Math. Mech., 15 (1966), pp. 137–146.

[10] R. F. Streater and A. S. Wightman, *PCT, Spin and Statistics and All That*, Benjamin, New York, 1964.

[11] V. S. Vladimirov, *Construction of envelopes of holomorphy for regions of a special type and their application*, Trudy Mat. Inst. Steklov., 60 (1961), pp. 101–144. (Transl. in Amer. Math. Soc. Transl. (2), 48 (1966), pp. 107–150.)

[12] N. I. Muskhelishvili, *Singular Integral Equations*, P. N. Noordhoff, Groningen, The Netherlands, 1953.

[13] E. C. Titchmarsch, *Introduction to the Theory of Fourier Integrals*, 2nd ed., Oxford University Press, London, 1948.

[14] W. Cauer, *The Poisson integral for functions with positive real parts*, Bull. Amer. Math. Soc., 38 (1932), pp. 713–714.

[15] M. R. Wohlers and E. J. Beltrami, *Distribution theory as the basis of generalized passive network analysis*, IEEE Trans. Circuit Theory, CT-12 (1965), pp. 164–169.

[16] A. Zemanian, *Distribution Theory and Transform Analysis*, McGraw-Hill, New York, 1965.

[17] B. Noble, *Methods Based on the Wiener-Hopf Technique*, Pergamon Press, New York, 1958.

A GENERALIZED WEIERSTRASS TRANSFORMATION*

A. H. ZEMANIAN†

1. Introduction. The Weierstrass transformation [1, Chap. 8],

$$F(s) = \frac{1}{\sqrt{4\pi}} \int_{-\infty}^{\infty} f(\tau) \exp\,[-\,(s-\tau)^2/4]\,d\tau, \tag{1.1}$$

maps a suitably restricted function $f(\tau)$ into an analytic function on a strip $\sigma_1 < \operatorname{Re} s < \sigma_2$ in the complex s-plane. Other names for it are the Gauss transformation [2], the Gauss-Weierstrass transformation [3, p. 578], and the Hille transformation [4]. The objective of this work is to extend the Weierstrass transformation and a certain inversion formula [1, p. 191] to allow $f(\tau)$ to be a certain type of generalized function.

This work is a sequel to a previous one [5] in which the convolution transformation of Hirschman and Widder [1],

$$F(x) = \int_{-\infty}^{\infty} f(\tau)G(x-\tau)\,d\tau, \tag{1.2}$$

was extended to certain classes of generalized functions. Here, the (nonfinite) kernel is defined by

$$G(\tau) = \frac{1}{2\pi i}\int_{-i\infty}^{i\infty} \frac{e^{z\tau}}{E(z)}\,dz, \tag{1.3}$$

where

$$E(z) = e^{-cz^2+bz} \prod_{\nu=1}^{\infty}\left(1-\frac{z}{a_\nu}\right)e^{z/a_\nu}, \tag{1.4}$$

the c, b, and a_ν are real, $c \geqq 0$, $a_\nu \neq 0$, $|a_\mu| \to \infty$, and $\sum a_\nu^{-2} < \infty$. However, in [5] we treated only those kernels for which $c = 0$. On the other hand, if we substitute the previously neglected factor $\exp(-cz^2)$ in place of $E(z)$ in (1.3), we obtain

$$G(\tau) = \frac{e^{-\tau^2/4c}}{\sqrt{4\pi c}}.$$

It is no restriction to set $c = 1$, [1, p. 171]. By substituting the resulting

* Received by the editors June 3, 1966, and in revised form October 6, 1966. Contributed at the Symposium on "The Applications of Generalized Functions" sponsored by the Air Force Office of Scientific Research at the 1966 Fall Meeting of Society for Industrial and Applied Mathematics held at the State University of New York at Stony Brook, September 12–14, 1966.

† Department of Applied Analysis, State University of New York at Stony Brook, Stony Brook, Long Island, New York 11790. This research was supported by the Air Force Cambridge Research Laboratories, Bedford, Massachusetts, under Contract AF19 (628)-2981.

$G(\tau)$ into (1.2) we obtain (1.1). It is in this fashion that the present paper complements our previous work. In §6, we combine the results of this paper with those of [5] to generalize the convolution transformation with kernels of the form (1.3) wherein c need not be zero.

As in [5], the basic idea here is to construct a testing function space $\mathcal{W}_{a,b}$, which contains the kernel of (1.1) for $a < \operatorname{Re} s < b$ and is closed under differentiation. The dual space $\mathcal{W}'_{a,b}$ is a space of generalized functions on which we define the generalized Weierstrass transformation by

$$F(s) = \frac{1}{\sqrt{4\pi}} \langle f(\tau), \exp[-(s-\tau)^2/4] \rangle, \qquad a < \operatorname{Re} s < b, \quad f \in \mathcal{W}'_{a,b}. \tag{1.5}$$

$F(s)$ turns out to be an analytic function for $a < \operatorname{Re} s < b$. Moreover, the Hirschman-Widder inversion formula [1, p. 191] is shown to be valid in terms of weak convergence in the distributional sense.

Our notation and terminology is the same as that of [5]. In particular, $\mathcal{R}$ denotes the real one-dimensional Euclidean space, and our testing functions are defined on $\mathcal{R}$. By a smooth function we mean a function that possesses continuous derivatives of all orders at all points of its domain. The notation $f(t)$ $(t \in \mathcal{R})$ for a generalized function f merely indicates that the testing functions on which f is defined have t as their independent variable. $\langle f, \phi \rangle$ denotes the number assigned to some ϕ in a testing function space by a member f of the dual space. The kth derivative of an ordinary or generalized function $f(t)$ is denoted alternatively by $d^k f/dt^k$, $D^k f$, $D_t^k f(t)$, or $f^{(k)}(t)$. Also, supp f denotes the support of f.

$\mathfrak{D}$ is the space of smooth functions on $\mathcal{R}$ having compact supports. The topology of $\mathfrak{D}$ is that which makes its dual the space $\mathfrak{D}'$ of Schwartz distributions on $\mathcal{R}$ [6, vol. I, p. 65]. $\mathcal{E}$ and $\mathcal{E}'$ are respectively the space of smooth functions on $\mathcal{R}$ and the dual space of distributions of compact support in $\mathcal{R}$. These spaces have their customary topologies [6, vol. I, pp. 88–90].

With s being the complex variable, $s = \sigma + i\omega$, and for $0 < t < \infty$, we set

$$k(s, t) = \frac{e^{-s^2/4t}}{\sqrt{4\pi t}}. \tag{1.6}$$

The kernel of the Weierstrass transformation (1.1) is $k(s - \tau, 1)$. Some computation establishes the following identity, wherein $P_n(s, 1/t)$ is a polynomial of degree n in both s and $1/t$ separately:

$$D_\tau^n k(s - \tau, t) = \frac{e^{-(s-\tau)^2/4t}}{\sqrt{4\pi t}} P_n\left(s - \tau, \frac{1}{\tau}\right). \tag{1.7}$$

2. The testing function space $\mathcal{W}_{a,b}$ and its dual $\mathcal{W}'_{a,b}$. Let a and b be fixed numbers in $\mathcal{R}$ with $a < b$. (That $a < b$ is always understood henceforth.) Let $\rho_{a,b}(\tau)$ be a smooth positive (never zero) function such that

$$\rho_{a,b}(\tau) = \begin{cases} e^{-b\tau/2}, & 1 < \tau < \infty, \\ e^{-a\tau/2}, & -\infty < \tau < -1. \end{cases}$$

It is also understood that $\rho_{a,b}(\tau)$ is a fixed, although unspecified, function on $-1 < \tau < 1$. $\mathcal{W}_{a,b}$ is the linear space of all complex-valued smooth functions $\phi(\tau)$ on $\mathcal{R}$ such that, for each $n = 0, 1, 2, \cdots$,

$$(2.1) \qquad \gamma_n(\phi) = \gamma_{a,b,n}(\phi) = \sup_{-\infty<\tau<\infty} |e^{\tau^2/4}\rho_{a,b}(\tau)\phi^{(n)}(\tau)| < \infty.$$

We assign to $\mathcal{W}_{a,b}$ the topology generated by the set of seminorms $\{\gamma_n\}$. Since γ_0 is a norm, $\mathcal{W}_{a,b}$ is a Hausdorff space. It is readily shown that $\mathcal{W}_{a,b}$ is sequentially complete and that differentiation is a continuous linear mapping of $\mathcal{W}_{a,b}$ into itself.

For any $\phi \in \mathcal{W}_{a,b}$, set

$$\gamma_n'(\phi) = \max_{0 \leqq \mu \leqq n} \gamma_\mu(\phi), \qquad n = 0, 1, 2, \cdots.$$

The set of norms $\{\gamma_n'\}$ generates a topology in $\mathcal{W}_{a,b}$ that is equivalent to that generated by $\{\gamma_n\}$. For every choice of p and q, the norms γ_p' and γ_q' are in concordance. This is proven in the same way as is the corresponding property for the space $\mathcal{L}_{a,b}$ defined in [5]. It now follows that $\mathcal{W}_{a,b}$ is a countably normed space [7, p. 6].

The dual $\mathcal{W}_{a,b}'$ of $\mathcal{W}_{a,b}$ is a linear space (under the customary definitions), which is sequentially complete with respect to both its strong and weak topologies [7, pp. 12–13]. For $f \in \mathcal{W}_{a,b}'$ and $\phi \in \mathcal{W}_{a,b}$, the derivative $f^{(1)}$ of f is defined by $\langle f^{(1)}, \phi\rangle = \langle f, -\phi^{(1)}\rangle$. From this it follows that differentiation is a continuous linear mapping of $\mathcal{W}_{a,b}'$ into $\mathcal{W}_{a,b}'$.

Some other properties of these spaces are the following:

1. Let $k(s, t)$ be defined by (1.6), and let s be a fixed complex number such that $a < \operatorname{Re} s < b$. Then, $k(s - \tau, 1)$ as a function of τ is a member of $\mathcal{W}_{a,b}$. This follows readily from (1.7).

2. For every fixed complex number s and real number t such that $0 < t < 1$, $k(s - \tau, t)$ as a function of τ is in $\mathcal{W}_{a,b}$. This also follows from (1.7).

3. For any given $f \in \mathcal{W}_{a,b}'$, there exist a positive constant C and a nonnegative integer r such that, for all $\phi \in \mathcal{W}_{a,b}$, $|\langle f, \phi\rangle| \leqq C\gamma_r'(\phi)$. This is proven in exactly the same way as is Theorem 3.3-1 of [8].

4. $\mathfrak{D}$ is a subset of $\mathcal{W}_{a,b}$, and the topology of $\mathfrak{D}$ is stronger than that induced on it by $\mathcal{W}_{a,b}$. Therefore, the restriction of $f \in \mathcal{W}_{a,b}'$ to $\mathfrak{D}$ is in $\mathfrak{D}'$.

5. $\mathcal{W}_{a,b}$ is a dense subset of the space $\mathcal{E}$, and the topology of $\mathcal{W}_{a,b}$ is

stronger than that induced on it by $\mathcal{E}$. Because of this, $\mathcal{E}'$ can be identified with a subset of $\mathcal{W}'_{a,b}$.

6. If $c \leqq a$ and $b \leqq d$, then $\mathcal{W}_{a,b}$ is a subset of $\mathcal{W}_{c,d}$, and the topology of $\mathcal{W}_{a,b}$ is stronger than that induced on it by $\mathcal{W}_{c,d}$. To see this, note that $\rho_{c,d}(\tau)/\rho_{a,b}(\tau)$ is a bounded function on $-\infty < \tau < \infty$. Denote its supremum by B. Then, for every $\phi \in \mathcal{W}_{a,b}$, $\gamma_{c,d,n}(\phi) \leqq B\gamma_{a,b,n}(\phi)$, and this verifies our assertion. It now follows that the restriction of $f \in \mathcal{W}'_{c,d}$ to $\mathcal{W}_{a,b}$ is in $\mathcal{W}'_{a,b}$.

7. Assume that f is a member of $\mathcal{W}'_{a,b}$ and also of $\mathcal{W}'_{c,d}$, where $a < b < c < d$. By this we mean that f is uniquely specified on $\mathcal{W}_{a,b} \cup \mathcal{W}_{c,d}$ (and therefore on $\mathfrak{D}$) and that the restrictions of f to $\mathcal{W}_{a,b}$ and $\mathcal{W}_{c,d}$ are in $\mathcal{W}'_{a,b}$ and $\mathcal{W}'_{c,d}$, respectively. We shall show here that there exists a unique member of $\mathcal{W}'_{a,d}$ whose restriction to $\mathcal{W}_{a,b} \cup \mathcal{W}_{c,d}$ coincides with f.

Let $\lambda(\tau)$ be a fixed smooth function such that $\lambda(\tau) = 0$ for $\tau < \tau_1$ and $\lambda(\tau) = 1$ for $\tau > \tau_2$, $-\infty < \tau_1 < \tau_2 < \infty$. Let us extend the definition of f onto the space $\mathcal{W}_{a,d}$ by means of the equation:

$$\langle f, \psi \rangle = \langle f, \lambda\psi \rangle + \langle f, (1 - \lambda)\psi \rangle, \qquad \psi \in \mathcal{W}_{a,d}.$$

The right-hand side has a sense since $\lambda\psi \in \mathcal{W}_{c,d}$ and $(1 - \lambda)\psi \in \mathcal{W}_{a,b}$.

This extension of f onto $\mathcal{W}_{a,d}$ is clearly linear. To show that it is continuous, invoke note 3:

$$|\langle f, \lambda\psi \rangle| \leqq C \max_{0 \leqq \mu \leqq n} \sup_{\tau} |e^{\tau^2/4}\rho_{c,d}(\tau)D^{\mu}[\lambda(\tau)\psi(\tau)]|$$

$$\leqq C \max_{0 \leqq \mu \leqq n} \sum_{p=0}^{\mu} \binom{\mu}{p} \sup_{\tau} \left| \left(\frac{\rho_{c,d}D^{\mu-p}\lambda}{\rho_{a,d}} \right) (e^{\tau^2/4}\rho_{a,d}D^{p}\psi) \right|.$$

Now, $|\rho_{c,d}\rho_{a,d}^{-1}D^{\mu-p}\lambda|$ is bounded on $-\infty < \tau < \infty$. Consequently, $\langle f, \lambda\psi_\nu \rangle \to 0$ as $\nu \to \infty$ whenever $\psi_\nu \to 0$ in $\mathcal{W}_{a,d}$. After similar considerations for $\langle f, (1 - \lambda)\psi \rangle$, we see that the above definition truly extends f into a member of $\mathcal{W}'_{a,d}$.

Finally, this extension of f into a member of $\mathcal{W}'_{a,d}$ is unique because, in view of the decomposition $\psi = \lambda\psi + (1 - \lambda)\psi$ for arbitrary $\psi \in \mathcal{W}_{a,d}$ and the facts that $\mathcal{W}_{a,b} \subset \mathcal{W}_{a,d}$ and $\mathcal{W}_{c,d} \subset \mathcal{W}_{a,d}$, there cannot be another member of $\mathcal{W}'_{a,d}$ having the same restriction to $\mathcal{W}_{a,b} \cup \mathcal{W}_{c,d}$.

8. Let $a < c$ and $d < b$. In place of the customary topology, assign to $\mathcal{W}_{c,d}$ the topology induced on it by $\mathcal{W}_{a,b}$. In this case, $\mathfrak{D}$ is dense in $\mathcal{W}_{c,d}$. Indeed, for any $\phi \in \mathcal{W}_{c,d}$, $\rho_{a,b}(\tau)D^k\phi(\tau) \to 0$ as $\tau \to \pm\infty$. Now, let $\lambda(\tau) \in \mathfrak{D}$ be such that $\lambda(\tau) = 1$ for $|\tau| < 1$ and $\lambda(\tau) = 0$ for $|\tau| > 2$. Also, let $\nu > 1$. Then, some straightforward computation shows that

$$\rho_{a,b}(\tau)D^k[\lambda(\tau/\nu)\phi(\tau) - \phi(\tau)] \to 0, \qquad \nu \to \infty.$$

Thus, $\{\lambda(\tau/\nu)\phi(\tau)\}_{\nu=1}^{\infty}$ is a sequence of elements in $\mathfrak{D}$ which converges in the induced topology of $\mathfrak{W}_{c,d}$ to $\phi(\tau)$. This establishes our assertion.

As a consequence, if $f \in \mathfrak{W}'_{a,b}$, then a knowledge of the values that f assigns to $\mathfrak{D}$ uniquely determines the values that f assigns to $\mathfrak{W}'_{c,d}$.

9. A locally Lebesgue-integrable function $h(\tau)$ which is such that $h(\tau)/e^{\tau^2/4}\rho_{a,b}(\tau)$ is absolutely integrable on $-\infty < \tau < \infty$ generates a member h of $\mathfrak{W}'_{a,b}$ through the definition:

$$\langle h, \phi\rangle = \int_{-\infty}^{\infty} h(\tau)\phi(\tau)\, d\tau, \qquad \phi \in \mathfrak{W}_{a,b}.$$

For,

$$|\langle h, \phi\rangle| = \left|\int_{-\infty}^{\infty} \frac{h(\tau)}{e^{\tau^2}\rho_{a,b}(\tau)}\, e^{\tau^2}\rho_{a,b}(\tau)\phi(\tau)\, d\tau\right.$$

$$\leqq \gamma_0(\phi)\int_{-\infty}^{\infty} \frac{|h(\tau)|}{e^{\tau^2}\rho_{a,b}(\tau)}\, d\tau.$$

As a consequence of this, the ordinary Weierstrass transform of h agrees with the generalized Weierstrass transform of h defined in the next section.

We end this section by establishing a "Fubini-type" lemma.

LEMMA 2.1. *Let $f(\tau) \in \mathfrak{W}'_{a,b}$, let $\psi(x)$ be a smooth function on the finite interval $A \leqq x \leqq B$, and let $\theta(\tau, x)$ be a smooth function on the domain, $-\infty < \tau < \infty$, $A \leqq x \leqq B$, such that, for every nonnegative integer n,*

$$\lim_{|\tau|\to\infty} e^{\tau^2/4}\rho_{a,b}(\tau)D_\tau^{\,n}\theta(\tau, x) = 0 \tag{2.2}$$

uniformly for $A \leqq x \leqq B$.

Then, the following three assertions are true:

(a) $\displaystyle\int_A^B \psi(x)\theta(\tau, x)\, dx \in \mathfrak{W}_{a,b}$,

(b) $\langle f(\tau), \theta(\tau, x)\rangle$ *is a continuous function on* $A \leqq x \leqq B$,

$$\text{(c)} \left\langle f(\tau), \int_A^B \psi(x)\theta(\tau, x)\, dx\right\rangle = \int_A^B \psi(x)\langle f(\tau), \theta(\tau, x)\rangle\, dx. \tag{2.3}$$

Proof. First note that

$$I(\tau) = \int_A^B \psi(x)\theta(\tau, x)\, dx$$

is a smooth function of τ and may be differentiated under the integral sign any number of times. Moreover, for each $n = 0, 1, 2, \cdots$,

$$\gamma_n\left[\int_A^B \psi(x)\theta(\tau, x)\, dx\right] = \sup_{-\infty<\tau<\infty}\left| e^{\tau^2/4}\rho_{a,b}(\tau)\int_A^B \psi(x)D_\tau^n\theta(\tau, x)\, dx\right|$$

$$\leqq \int_A^B |\psi(x)|\, dx \sup_{-\infty<\tau<\infty}[e^{\tau^2/4}\rho_{a,b}(\tau) \sup_{A<x<B} |D_\tau^n\theta(\tau, x)|].$$

The last expression is finite because of (2.2), and this establishes (a).

To prove (b), fix x as a point in the closed interval $[A, B]$, and restrict Δx so that $x + \Delta x$ is also a point in $[A, B]$. We will have proven that $\langle f(\tau), \theta(\tau, x)\rangle$ is a continuous function on $A \leqq x \leqq B$ when we show that $\theta(\tau, x + \Delta x) - \theta(\tau, x) \to 0$ in $\mathcal{W}_{a,b}$ as $\Delta x \to 0$. In view of (2.2), given an $\epsilon > 0$ and a nonnegative integer n, we can choose T so large that, for all $|\tau| > T$ and all permissible Δx,

$$|e^{\tau^2/4}\rho_{a,b}(\tau)D_\tau^n[\theta(\tau, x + \Delta x) - \theta(\tau, x)]| < \frac{\epsilon}{2}. \tag{2.4}$$

Fix T this way. Since the left-hand side of (2.4) is a continuous function of $(\tau, \Delta x)$, it tends to zero as $\Delta x \to 0$ uniformly on $-T \leqq \tau \leqq T$. Thus, $\theta(\tau, x + \Delta x) - \theta(\tau, x)$ does tend to zero in $\mathcal{W}_{a,b}$ as $\Delta x \to 0$.

We now prove assertion (c) by making use of certain "Riemann sums" for the integrals therein. Set $X = B - A$. If we can show that, as $m \to \infty$,

$$J(\tau, m) = \frac{X}{m}\sum_{\nu=1}^{m} \psi\left(A + \frac{\nu X}{m}\right)\theta\left(\tau, A + \frac{\nu X}{m}\right)$$

converges in $\mathcal{W}_{a,b}$ to $I(\tau)$, then we can write

$$\left\langle f(\tau), \frac{X}{m}\sum_{\nu=1}^{m} \psi\left(A + \frac{\nu X}{m}\right)\theta\left(\tau, A + \frac{\nu X}{m}\right)\right\rangle \tag{2.5}$$
$$\to \left\langle f(\tau), \int_A^B \psi(x)\theta(\tau, x)\, dx\right\rangle, \quad m \to \infty.$$

On the other hand, we may apply $f(\tau)$ term by term to the sum in the left-hand side of (2.5), and, in view of (b), the result tends to

$$\int_A^B \psi(x)\langle f(\tau), \theta(\tau, x)\rangle\, dx$$

as $m \to \infty$. Assertion (c) will thereby be proven.

Throughout the following, n is fixed, and $\psi(x) \not\equiv 0$. Set

$$H(\tau, m) = e^{\tau^2/4}\rho_{a,b}(\tau)D_\tau^n[I(\tau) - J(\tau, m)].$$

We have to show that $H(\tau, m)$ tends to zero as $m \to \infty$ uniformly on $-\infty < \tau < \infty$.

In view of (2.2), given an $\epsilon > 0$, there exists a T such that, for $|\tau| > T$ and $A \leqq x \leqq B$,

$$| e^{\tau^2/4} \rho_{a,b}(\tau) D_\tau^n \theta(\tau, x)| < \frac{\epsilon}{3}\left[\int_A^B |\psi(x)|\, dx\right]^{-1}.$$

Consequently,

$$\sup_{|\tau|>T} | e^{\tau^2/4} \rho_{a,b}(\tau) D_\tau^n I(\tau)| < \frac{\epsilon}{3}. \tag{2.6}$$

Also, for all m,

$$\begin{aligned}&\sup_{|\tau|>T} | e^{\tau^2/4} \rho_{a,b}(\tau) D_\tau^n J(\tau, m)| \\ &\qquad < \frac{\epsilon}{3}\left[\int_A^B |\psi(x)|\, dx\right]^{-1} \frac{X}{m} \sum_{\nu=1}^{m} \left|\psi\left(A + \frac{\nu X}{m}\right)\right|.\end{aligned} \tag{2.7}$$

Moreover, there exists an m_0 such that for all $m > m_0$ the right-hand side of (2.7) is bounded by $2\epsilon/3$. Thus, for $m > m_0$ and $|\tau| > T$, $|H(\tau, m)| < \epsilon$.

Having fixed T this way, set

$$K = \sup_{|\tau|<T} | e^{\tau^2/4} \rho_{a,b}(\tau)|.$$

Then, for $|\tau| \leqq T$,

$$\begin{aligned}| H(\tau, m)| \leqq K \Bigg| &\int_A^B \psi(x) D_\tau^n \theta(\tau, x)\, dx \\ &- \frac{X}{m} \sum_{\nu=1}^{m} \psi\left(A + \frac{\nu X}{m}\right) D_\tau^n \theta\left(\tau, A + \frac{\nu X}{m}\right)\Bigg|.\end{aligned} \tag{2.8}$$

Since $\psi(x) D_\tau^n \theta(\tau, x)$ is uniformly continuous for all (τ, x) such that $-T \leqq \tau \leqq T$ and $A \leqq x \leqq B$, there exists an m_1 such that for all $m > m_1$ the right-hand side of (2.8) is bounded by ϵ on $-T \leqq \tau \leqq T$. This completes the proof.

3. The generalized Weierstrass transformation. We shall call f a Weierstrass-transformable generalized function if it possesses the following properties: (i) f is a functional on some domain $d(f)$ of functions; (ii) f is additive, that is, if $\phi, \theta, \phi + \theta \in d(f)$, then $\langle f, \phi + \theta\rangle = \langle f, \phi\rangle + \langle f, \theta\rangle$; (iii) $\mathcal{W}_{a,b} \subset d(f)$ for at least one pair of real numbers a and b with $a < b$; (iv) for every $\mathcal{W}_{a,b} \subset d(f)$, the restriction of f to $\mathcal{W}_{a,b}$ is in $\mathcal{W}'_{a,b}$.

Corresponding to f there exists a unique set Λ_f on the real line defined as follows: A point σ is in Λ_f if and only if there exist two real numbers a_σ, b_σ with

$a_\sigma < \sigma < b_\sigma$, $\mathfrak{W}_{a_\sigma,b_\sigma} \subset d(f)$. Let σ_1 be the infimum of Λ_f and σ_2 the supremum of Λ_f. (Possibly, $\sigma_1 = -\infty$ and $\sigma_2 = +\infty$.) Consequently, Λ_f is the union of all open intervals (a_σ, b_σ) arising as σ varies through Λ_f and, for each σ, (a_σ, b_σ) varies through all choices for which $\mathfrak{W}_{a_\sigma,b_\sigma} \subset d(f)$. Therefore, Λ_f is an open set. It also follows that there exist points a, b, c, and d in Λ_f, with $a < b$, $c < d$, a arbitrarily close to σ_1, and d arbitrarily close to σ_2, such that $f \in \mathfrak{W}'_{a,b}$ and $f \in \mathfrak{W}'_{c,d}$. But then, by note 7 in §2, f possesses a unique extension into a member of $\mathfrak{W}'_{a,d}$. Consequently, there exists a functional f_1, which is unique on every $\mathfrak{W}_{\alpha,\beta}$ for which $\alpha > \sigma_1$ and $\beta < \sigma_2$ $(\alpha < \beta)$, with the following two properties: (a) the restriction of f_1 to any such $\mathfrak{W}_{\alpha,\beta}$ is in $\mathfrak{W}'_{\alpha,\beta}$; (b) the restriction of f_1 to any $\mathfrak{W}_{a,b} \subset d(f)$ coincides with the original f. (The assertion (b) follows from the additivity property of f and a decomposition of any $\phi \in \mathfrak{W}_{a,b}$ in accordance with note 7 of §2.)

By the above definition, f_1 is also Weierstrass-transformable and the corresponding set Λ_{f_1} is the open interval $\sigma_1 < \sigma < \sigma_2$. Henceforth, it is understood that every Weierstrass-transformable generalized function f has been extended into the functional f_1, which we again denote by f.

Incidentally, under this convention, every Weierstrass-transformable generalized function f is a continuous linear functional on the union space $\mathfrak{W}(\sigma_1, \sigma_2)$ of all $\mathfrak{W}_{a,b}$ spaces for which $[a, b] \subset (\sigma_1, \sigma_2)$. (See [7, pp. 22–24].) This is a consequence of note 6 in §2.

We are at last ready to define the Weierstrass transform $F(s)$ of a Weierstrass-transformable generalized function $f(\tau)$. By note 1 of §2, for each fixed s in the strip $\Omega_f = \{s : \sigma_1 < \operatorname{Re} s < \sigma_2\}$, the function $k(s - \tau, 1)$ as a function of τ is in $\mathfrak{W}_{c,d}$, where we now choose c and d such that $\sigma_1 < c < \operatorname{Re} s < d < \sigma_2$. We define $F(s)$ as the application of $f \in \mathfrak{W}_{c,d}$ to $k(s - \tau, 1) \in \mathfrak{W}_{c,d}$:

$$F(s) = \langle f(\tau), k(s - \tau, 1)\rangle, \qquad s \in \Omega_f. \tag{3.1}$$

Ω_f will be called the strip of definition for (3.1) and is determined by f via $\Lambda_f : \Omega_f = \{s : \operatorname{Re} s \in \Lambda_f\}$.

THEOREM 3.1. *Let f be a Weierstrass-transformable generalized function with the strip of definition $\Omega_f = \{s : \sigma_1 < \operatorname{Re} s < \sigma_2\}$, and let $F(s)$ be defined by (3.1). Then, $F(s)$ is an analytic function on Ω_f, and*

$$\begin{gathered} F^{(n)}(s) = \langle f(\tau), D_s^n k(s - \tau, 1)\rangle, \\ D_s = \frac{\partial}{\partial s}, \quad s \in \Omega_f, \quad n = 1, 2, 3, \cdots. \end{gathered} \tag{3.2}$$

Moreover, on any closed substrip $a \leqq \operatorname{Re} s = \sigma \leqq b$ of Ω_f (i.e., $\sigma_1 < a < b < \sigma_2$), we have

$$|F(\sigma + i\omega)| \leqq e^{\omega^2/4} B(|\omega|), \tag{3.3}$$

where B is a polynomial which depends on f and the choice of the substrip $a \leqq \operatorname{Re} s \leqq b$.

Proof. We prove (3.2) through an inductive argument. Assume that (3.2) is true for n replaced by $n - 1$. It is true by definition for $n = 0$. Let s be an arbitrary but fixed point in Ω_f, and choose a and b such that $\sigma_1 < a < \operatorname{Re} s < b < \sigma_2$. Therefore, $f \in \mathcal{W}'_{a,b}$. Also, let $r > 0$ be such that $a < \operatorname{Re} s - r < \operatorname{Re} s + r < b$, and let $|\Delta s| < r$. For $\Delta s \neq 0$, consider the equation

$$\begin{aligned}(3.4)\qquad &\frac{1}{\Delta s}[F^{(n-1)}(s + \Delta s) - F^{(n-1)}(s)] \\ &\qquad - \langle f(\tau), D_s^n k(s - \tau, 1)\rangle = \langle f(\tau), \psi_{\Delta s}(\tau)\rangle,\end{aligned}$$

where

$$\psi_{\Delta s}(\tau) = \frac{1}{\Delta s}[D_s^{n-1}k(s + \Delta s - \tau, 1) - D_s^{n-1}k(s - \tau, 1)] - D_s^n k(s - \tau, 1).$$

For $\Delta s = 0$, set $\psi_{\Delta s}(\tau) \equiv 0$. We have noted in §2 that $k(s - \tau, 1) \in \mathcal{W}_{a,b}$ and that $\mathcal{W}_{a,b}$ is closed under differentiation. It follows that (3.4) has a sense. We shall prove that $F(s)$ is analytic on Ω_f and satisfies (3.2) by showing that $\psi_{\Delta s}(\tau) \to 0$ in $\mathcal{W}_{a,b}$ as $|\Delta s| \to 0$.

By using (1.7) and the fact that $D_s k(s - \tau, 1) = -D_\tau k(s - \tau, 1)$, we may write, for any nonnegative integer m,

$$\begin{aligned}(3.5)\qquad \psi_{\Delta s}^{(m)}(\tau) &= \frac{1}{\Delta s}[k(s + \Delta s - \tau, 1)Q(s + \Delta s - \tau) \\ &\qquad - k(s - \tau, 1)Q(s - \tau)] - D_s[k(s - \tau, 1)Q(s - \tau)],\end{aligned}$$

where Q is a polynomial. Let C be a circle with center at s and radius r_1, where $a < \operatorname{Re} s - r_1 < \operatorname{Re} s - r < \operatorname{Re} s + r < \operatorname{Re} s + r_1 < b$. Since the right-hand side of (3.5) is an entire function of s, we may invoke Cauchy's integral formula to write

$$\begin{aligned}\psi_{\Delta s}^{(m)}(\tau) &= \frac{1}{\Delta s 2\pi i}\int_C k(\zeta - \tau, 1)Q(\zeta - \tau)\left[\frac{1}{\zeta - s - \Delta s} - \frac{1}{\zeta - s}\right] d\zeta \\ &\qquad - \frac{1}{2\pi i}\int_C \frac{k(\zeta - \tau, 1)Q(\zeta - \tau)}{(\zeta - s)^2}\, d\zeta.\end{aligned}$$

Hence,

$$e^{\tau^2/4}\rho_{a,b}(\tau)\psi_{\Delta s}^{(m)}(\tau) = \frac{\Delta s}{2\pi i}\int_C \frac{A(\zeta, \tau)}{(\zeta - s - \Delta s)(\zeta - s)^2}\, d\zeta,$$

where

$$A(\zeta, \tau) = e^{\tau^2/4}\rho_{a,b}(\tau)k(\zeta - \tau, 1)Q(\zeta - \tau)$$
$$= (4\pi)^{-1/2}e^{-\zeta^2/4}\rho_{a,b}(\tau)e^{\zeta\tau/2}Q(\zeta - \tau).$$

Since ζ is restricted to the circle C, which is contained in the strip $a < \text{Re}\, s < b$, we see from the definition of $\rho_{a,b}(\tau)$ that $|A(\zeta, \tau)|$ is bounded on the domain $-\infty < \tau < \infty$, $\zeta \in C$ by, say, the constant M. Therefore,

$$|e^{\tau^2/4}\rho_{a,b}(\tau)\psi_{\Delta s}^{(m)}(\tau)| \leqq \frac{M|\Delta s|}{2\pi}\int_C \frac{|d\zeta|}{(r_1 - r)r_1^2} = |\Delta s|\frac{M}{(r_1 - r)r_1}.$$

This shows that as $|\Delta s| \to 0$ the left-hand side converges to zero uniformly on $-\infty < \tau < \infty$, which is what we wished to show.

We turn now to the proof of (3.3). By note 3 of §2 and (1.7), there exist a positive constant C and a nonnegative integer r such that, for $\sigma_1 < c < a \leqq \sigma \leqq b < d < \sigma_2$,

$$|F(\sigma + i\omega)| \leqq C \max_{0 \leqq n \leqq r} \sup_{-\infty<\tau<\infty} |e^{\tau^2/4}\rho_{c,d}(\tau)D_\tau^n k(\sigma + i\omega - \tau, 1)|$$

$$= \frac{C}{\sqrt{4\pi}}\exp[(\omega^2 - \sigma^2)/4] \max_{0 \leqq n \leqq r} \sup_{-\infty<\tau<\infty} |e^{\sigma\tau/2}\rho_{c,d}(\tau)P_n(\sigma + i\omega - \tau, 1)|.$$

In view of the definition of $\rho_{c,d}(\tau)$, the last expression implies (3.3). That $B(|\omega|)$ depends in general on the choice of the strip $a \leqq \text{Re}\, s \leqq b$ is indicated by the example [1, p. 178]:

$$\langle e^{\tau^2/5}, k(s - \tau, 1)\rangle = \sqrt{5}\, e^{s^2}, \qquad -\infty < \text{Re}\, s < \infty$$

This completes the proof of Theorem 3.1.

4. Inversion. The Hirschman-Widder inversion formula for the ordinary Weierstrass transformation is the following: Let $F(s)$ be defined by (1.1) and $k(s, t)$ by (1.6). Then, under certain conditions on the function f, we have

$$(4.1) \quad f(\tau) = \lim_{t \to 1-} \int_{-\infty}^{\infty} k(\omega + i\tau - i\sigma, t)F(\sigma + i\omega)\, d\omega, \qquad \sigma_1 < \sigma < \sigma_2.$$

The objective of this section is to prove that this formula also inverts our generalized Weierstrass transformation (3.1), when the limit as $t \to 1-$ is taken in the sense of weak convergence in $\mathfrak{D}'$.

THEOREM 4.1. *Let f be a Weierstrass-transformable generalized function with the strip of definition $\Omega_f = \{s{:}\sigma_1 < \text{Re}\, s < \sigma_2\}$, let $F(s)$ be defined by (3.1), and let σ be any fixed real number such that $\sigma_1 < \sigma < \sigma_2$. Then, for each*

$\phi \in \mathfrak{D}$,

$$(4.2)\quad \lim_{t\to 1-} \left\langle \int_{-\infty}^{\infty} k(\omega + i\tau - i\sigma, t)F(\sigma + i\omega)\, d\omega, \phi(\tau) \right\rangle = \langle f(\tau), \phi(\tau)\rangle.$$

Remark. In view of note 8 of §2, the inversion formula (4.2) uniquely determines f as a member of $\mathcal{W}'_{c,d}$ whenever $\sigma_1 < c < d < \sigma_2$. In other words, this is a complete inversion in the sense that f is determined as a member of the dual of the union space $\mathcal{W}(\sigma_1, \sigma_2)$ of all $\mathcal{W}_{c,d}$ spaces for which $[c, d] \subset (\sigma_1, \sigma_2)$.

Furthermore, Theorems 3.1 and 4.1 show that the generalized Weierstrass transformation is a one-to-one mapping from the dual of $\mathcal{W}(\sigma_1, \sigma_2)$ into a set of analytic functions possessing the property asserted by the last sentence of Theorem 3.1.

Proof of Theorem 4.1. Formally, the proof runs as follows. For $0 < t < 1$ and for $\sigma_1 < a < \sigma < b < \sigma_2$,

$$(4.3)\quad \left\langle \int_{-\infty}^{\infty} k(\omega + ix - i\sigma, t)F(\sigma + i\omega)\, d\omega, \phi(x) \right\rangle$$

$$(4.4)\quad = \left\langle \int_{-\infty}^{\infty} k(\omega + ix - i\sigma, t)\langle f(\tau), k(\sigma + i\omega - \tau, 1)\rangle\, d\omega, \phi(x) \right\rangle$$

$$(4.5)\quad = \left\langle \left\langle f(\tau), \int_{-\infty}^{\infty} k(\omega + ix - i\sigma, t)k(\sigma + i\omega - \tau, 1)\, d\omega \right\rangle, \phi(x) \right\rangle$$

$$(4.6)\quad = \langle \phi(x), \langle f(\tau), k(x - \tau, 1 - t)\rangle\rangle$$

$$(4.7)\quad = \langle f(\tau), \langle \phi(x), k(x - \tau, 1 - t)\rangle\rangle.$$

The proof is completed by showing that $\langle \phi(x), k(x - \tau, 1 - t)\rangle \to \phi(\tau)$ in $\mathcal{W}_{a,b}$ as $t \to 1-$.

We now justify the various steps of this formal argument. By Theorem 3.1, for s restricted to any substrip $a \leqq \operatorname{Re} s = \sigma \leqq b$ of Ω_f ($\sigma_1 < a$, $b < \sigma_2$), we have

$$(4.8)\quad | k(\omega + ix - i\sigma, t)F(\sigma + i\omega) | \leqq \frac{e^{(x-\sigma)^2/4t}}{\sqrt{4\pi t}}\, e^{(1/4)\omega^2(1-1/t)} B(| \omega |).$$

Since $0 < t < 1$, this shows that

$$(4.9)\quad \int_{-\infty}^{\infty} k(\omega + ix - i\sigma, t)F(\sigma + i\omega)\, d\omega$$

converges uniformly for x restricted to any compact set, which implies that (4.9) is a continuous function of x. (By Cauchy's theorem, this also shows that we may change the value of σ in the range $\sigma_1 < \sigma < \sigma_2$ without altering

(4.9).) Thus, (4.3) and therefore (4.4) have a sense as the integral of the product of (4.9) and $\phi(x)$.

Next, we shall prove that (4.4) is equal to (4.5). We shall need the following lemma.

LEMMA 4.2. *Let f be defined as in Theorem 4.1, and let x, t, and σ be fixed, $0 < t < 1$, $\sigma_1 < \sigma < \sigma_2$. Given any $\epsilon > 0$, the positive numbers ω_1 and ω_2 can be so chosen that*

$$\left|\left(\int_{-\infty}^{-\omega_1} + \int_{\omega_2}^{\infty}\right) k(\omega + ix - i\sigma, t) \langle f(\tau), k(\sigma + i\omega - \tau, 1)\rangle \, d\omega\right| < \epsilon, \tag{4.10}$$

and simultaneously,

$$\left|\left\langle f(\tau), \left(\int_{-\infty}^{-\omega_1} + \int_{\omega_2}^{\infty}\right) k(\omega + ix - i\sigma, t) k(\sigma + i\omega - \tau, 1) \, d\omega\right\rangle\right| < \epsilon. \tag{4.11}$$

Proof. The inequality (4.10) follows immediately from (4.8). To establish (4.11), first note that we may differentiate

$$\left(\int_{-\infty}^{-\omega_1} + \int_{\omega_2}^{\infty}\right) k(\omega + ix - i\sigma, t) k(\sigma + i\omega - \tau, 1) \, d\omega \tag{4.12}$$

with respect to τ under the integral sign any number of times. This is because, if τ is confined to any compact set, (1.7) implies that

$$| k(\omega + ix - i\sigma, t) D_\tau^n k(\sigma + i\omega - \tau, 1) | \leqq e^{(1/4)\omega^2(1-1/t)} Q(|\omega|),$$

where Q is a polynomial. Thus, every differentiation with respect to τ of (4.12) under the integral sign leads to an integral which converges uniformly on every bounded τ interval.

By differentiating (4.12) with respect to τ under the integral sign n times and by again using (1.7), we obtain

$$\begin{aligned} &\left| e^{\tau^2/4} \rho_{a,b}(\tau) D_\tau^n \left(\int_{-\infty}^{-\omega_1} + \int_{\omega_2}^{\infty}\right) k(\omega + ix - i\sigma, t) k(\sigma + i\omega - \tau, 1) \, d\tau \right| \\ &\qquad \leqq \sum_\nu{}^{*} e^{\sigma\tau/2} \rho_{a,b}(\tau) R_\nu(|\tau|) \left(\int_{-\infty}^{-\omega_1} + \int_{\omega_2}^{\infty}\right) e^{(1/4)\omega^2(1-1/t)} S_\nu(|\omega|) \, d\omega. \end{aligned} \tag{4.13}$$

Here $\sum^*$ denotes a finite sum, and R_ν and S_ν denote monomials. Choose a and b such that $\sigma_1 < a < \sigma < b < \sigma_2$. Then, $e^{\sigma\tau/2}\rho_{a,b}(\tau)R_\nu(|\tau|)$ is bounded on $-\infty < \tau < \infty$. Since $0 < t < 1$, the right-hand side of (4.13) is also a bounded function of τ, which proves that (4.12) is a member of $\mathcal{W}_{a,b}$. Moreover, the right-hand side of (4.13) can be made arbitrarily small by choosing ω_1 and ω_2 sufficiently large. In view of note 3 in §2, this verifies (4.11) and completes the proof of Lemma 4.2.

Next, by (1.7) once again, we see that

$$| e^{\tau^2/4}\rho_{a,b}(\tau)D_\tau^{\,n}k(\sigma + i\omega - \tau, 1) |$$
$$= \frac{1}{\sqrt{4\pi}} e^{(\omega^2-\sigma^2)/4}\rho_{a,b}(\tau)e^{\sigma\tau/2} \,| P_n(\sigma + i\omega - \tau, 1)|,$$

and, if $a < \sigma < b$, this quantity tends to zero as $|\tau| \to \infty$ uniformly on $-\omega_1 \leqq \omega \leqq \omega_2$. Thus, from Lemma 2.1 we have that, for $\sigma_1 < a < \sigma < b < \sigma_2$ and any finite positive ω_1 and ω_2,

$$\tag{4.14} \begin{aligned} &\int_{-\omega_1}^{\omega_2} k(\omega + ix - i\sigma, t)\ \langle f(\tau), k(\sigma + i\omega - \tau, 1)\rangle\, d\omega \\ &= \left\langle f(\tau), \int_{-\omega_1}^{\omega_2} k(\omega + ix - i\sigma, t)k(\sigma + i\omega - \tau, 1)\, d\omega \right\rangle. \end{aligned}$$

By combining this equation with Lemma 4.2, we see that (4.4) is truly equal to (4.5) when $\sigma_1 < a < \sigma < b < \sigma_2$ and $0 < t < 1$.

That (4.5) is equal to (4.6) follows from a known result [1, Theorem 2.5, p. 177]. Next, by definition, the support of $\phi(x) \in \mathfrak{D}$ is contained in a finite interval, say, $A \leqq x \leqq B$. Moreover,

$$e^{\tau^2/4}\rho_{a,b}(\tau)D_\tau^{\,n}k(x - \tau, 1 - t)$$
$$= \frac{e^{-x^2/4(1-t)}}{\sqrt{4\pi(1 - t)}} \rho_{a,b}(\tau)e^{x\tau/2(1-t)}e^{(1/4)\tau^2[1-1/(1-t)]}P_n\left(x - \tau, \frac{1}{1-t}\right).$$

Since $1 - 1/(1 - t) < 0$, for every choice of a and b this quantity tends to zero as $|\tau| \to \infty$ uniformly on $A \leqq x \leqq B$. Therefore, we can invoke Lemma 2.1 again, which proves that (4.6) is equal to (4.7) when $\sigma_1 < a < b < \sigma_2$ and $0 < t < 1$.

As the last step, we prove that, as $t \to 1-$, $\langle \phi(x), k(x - \tau, 1 - t)\rangle \to \phi(\tau)$ in $\mathfrak{W}_{a,b}$ for every a and b, $a < b$. Set $x = \tau + 2y\sqrt{1 - t}$, where τ and t are fixed $(0 < t < 1)$. Then,

$$\tag{4.15} \frac{1}{\sqrt{4\pi(1 - t)}} \int_{-\infty}^{\infty} \exp \frac{(x - \tau)^2}{4(t - 1)}\, dx = \pi^{-1/2} \int_{-\infty}^{\infty} e^{-y^2}\, dy = 1.$$

In view of this equation and the fact that $\phi \in \mathfrak{D}$, we may differentiate under the integral sign and integrate by parts n times to get

$$\tag{4.16} \begin{aligned} &e^{\tau^2/4}\rho_{a,b}(\tau)\ D_\tau^{\,n}[\langle \phi(x), k(x - \tau, 1 - t)\rangle - \phi(\tau)] \\ &= \frac{e^{\tau^2/4}\rho_{a,b}(\tau)}{\sqrt{4\pi(1 - t)}} \int_{-\infty}^{\infty} [\phi^{(n)}(x) - \phi^{(n)}(\tau)] \exp \frac{(x - \tau)^2}{4(t - 1)}\, dx \\ &= I_1 + I_2 + I_3. \end{aligned}$$

Here, I_1, I_2, and I_3 denote respectively the expressions obtained after the integration on $-\infty < x < \infty$ is broken up into integrations on $-\infty < x < \tau - \delta$, $\tau - \delta \leqq x \leqq \tau + \delta$, and $\tau + \delta < x < \infty$ ($\delta > 0$).

Consider $I_2(\tau)$. In view of (4.15), we have

$$|I_2(\tau)| \leqq e^{\tau^2/4}\rho_{a,b}(\tau) \sup_{\tau-\delta<x<\tau+\delta} |\phi^{(n)}(x) - \phi^{(n)}(\tau)|$$
$$\leqq \delta e^{\tau^2/4}\rho_{a,b}(\tau) \sup_{\tau-\delta<y<\tau+\delta} |\phi^{(n+1)}(y)|.$$

Since $\phi \in \mathfrak{D}$, the right-hand side is dominated by δB, where B is a constant with respect to t, τ, and δ ($0 < t < 1$, $-\infty < \tau < \infty$, and $0 < \delta < 1$). Therefore, given an $\epsilon > 0$, we have $|I_2(\tau)| < \epsilon$ for $\delta = \min(1, \epsilon/B)$. Fix δ this way.

Next, consider

$$(4.17) \quad \begin{aligned} I_1(\tau) &= \frac{e^{\tau^2/4}\rho_{a,b}(\tau)}{\sqrt{4\pi(1-t)}} \int_{-\infty}^{\tau-\delta} \phi^{(n)}(x) \exp\frac{(x-\tau)^2}{4(t-1)}\,dx \\ &\quad - \frac{e^{\tau^2/4}\rho_{a,b}(\tau)}{\sqrt{4\pi(1-t)}} \phi^{(n)}(\tau) \int_{-\infty}^{\tau-\delta} \exp\frac{(x-\tau)^2}{4(t-1)}\,dx. \end{aligned}$$

By the change of variable $x = \tau + 2y\sqrt{1-t}$, we obtain

$$\frac{1}{\sqrt{4\pi(1-t)}} \int_{-\infty}^{\tau-\delta} \exp\frac{(x-\tau)^2}{4(t-1)}\,dx = \pi^{-1/2} \int_{-\infty}^{-\delta/2\sqrt{1-t}} e^{-y^2}\,dy \to 0, \qquad \tau \to 1-.$$

Since the support of ϕ is bounded, this shows that the second term on the right-hand side of (4.17) tends uniformly to zero as $t \to 1-$ on $-\infty < \tau < \infty$.

Let $J_1(\tau)$ denote the first term on the right-hand side of (4.17), and assume that the support of $\phi(x)$ is contained in the finite interval $A \leqq x \leqq B$. Then, for $-\infty < \tau - \delta < A$, $J_1(\tau) \equiv 0$. For $A \leqq \tau - \delta \leqq B$, let K be a (constant) bound on $e^{\tau^2/4}\rho_{a,b}(\tau)$. Therefore,

$$|J_1(\tau)| \leqq \frac{Ke^{\delta^2/4(t-1)}}{\sqrt{4\pi(1-t)}} \int_A^B |\phi^{(n)}(x)|\,dx \to 0, \qquad t \to 1-.$$

Hence, as $t \to 1-$, $J_1(\tau) \to 0$ uniformly on $A \leqq \tau - \delta \leqq B$.

Finally, consider the range $B < \tau - \delta < \infty$. There exists a constant M such that $\rho_{a,b}(\tau) < Me^{-b\tau/2}$ for $B < \tau - \delta < \infty$. Consequently, since $0 < t < 1$,

$$(4.18) \quad \begin{aligned} |J_1(\tau)| &\leqq \frac{M}{\sqrt{4\pi(1-t)}} \exp\left[\frac{\tau^2}{4} - \frac{b\tau}{2} + \frac{(B-\tau)^2}{4(t-1)}\right] \\ &\quad \cdot \int_A^B |\phi^{(n)}(x)|\,dx. \end{aligned}$$

Some computation shows that

$$(4.19) \qquad \exp\left[\frac{\tau^2}{4} - \frac{b\tau}{2} + \frac{(B - \tau)^2}{4(t - 1)}\right]$$

tends to zero as $|\tau| \to \infty$ and has a single maximum which occurs at the point

$$\tau_m = \frac{B}{t} + b\left(1 - \frac{1}{t}\right).$$

Therefore, there exists a t_1 such that, for $t_1 < t < 1$, the single maximum of (4.19) lies within the interval $(B - \delta, B + \delta)$. Since $B < \tau - \delta < \infty$, we may set $\tau = B + \delta$ to get, for $t_1 < t < 1$,

$$|J_1(\tau)|$$
$$\leqq M \exp\left[\frac{(B + \delta)^2}{4} - \frac{b(B + \delta)}{2}\right] \int_A^B |\phi^{(n)}(x)|\, dx \frac{e^{\delta^2/4(t-1)}}{\sqrt{4\pi(1 - t)}} \to 0,$$
$$t \to 1-.$$

Thus, as $t \to 1-$, $|J_1(\tau)|$ converges uniformly to zero on $B < \tau - \delta < \infty$ as well.

Altogether, we have demonstrated that, as $t \to 1-$, $I_1(\tau)$ converges to zero uniformly on $-\infty < \tau < \infty$. A similar argument shows that $I_3(\tau)$ also converges to zero uniformly on $-\infty < \tau < \infty$ as $t \to 1-$. In view of (4.16), this proves that, for each $n = 0, 1, 2, \cdots$,

$$\limsup_{t \to 1-} \gamma_n[\langle \phi(x), k(x - \tau, 1 - t)\rangle - \phi(\tau)] \leqq \epsilon.$$

Since ϵ is arbitrary, the proof of Theorem 4.1 is complete.

5. An application to the problem of heat flow. A physical interpretation can be assigned to the expression

$$(5.1) \qquad u(x, t) = \langle f(\tau), k(x - \tau, t)\rangle, \qquad 0 < t < 1.$$

Namely, $u(x, t)$ is the temperature in an infinitely long uniform rod at the instant t when the temperature at $t = 0$ is specified as the generalized function $f(x)$. More precisely, we can state the following theorem.

THEOREM 5.1. *If $f \in \mathcal{W}'_{a,b}$ for some a and b $(a < b)$ and if $u(x, t)$ is defined by* (5.1), *then on the domain $-\infty < x < \infty, 0 < t < 1$, $u(x, t)$ has the ordinary partial derivatives $D_x^2 u$ and $D_t u$, which satisfy $D_x^2 u = D_t u$. Moreover, as $t \to 0+$, $u(x, t)$ as a function of x converges weakly in $\mathcal{D}'$ to $f(x)$.*

Proof. We shall merely sketch out the proof of this theorem because its details are quite similar to what we have already presented. First of all, we can show that

$$(5.2) \qquad D_x^n u(x, t) = \frac{\partial^n}{\partial x^n} u(x, t) = \langle f(\tau), D_x^n k(x - \tau, t)\rangle, \qquad n = 1, 2, 3, \cdots,$$

through an argument that is almost identical to the proof of (3.2).

Secondly, we prove that

$$(5.3) \qquad D_t u(x, t) = \langle f(\tau), D_t k(x - \tau, t)\rangle, \qquad 0 < t < 1.$$

It is easy to show that, as a function of τ, $D_t k(x - \tau, t)$ is in $\mathfrak{W}_{a,b}$ for every a and $b (a < b)$, so that the right-hand side of (5.3) has a sense. Next, consider $\langle f(\tau), \theta_{\Delta t}(\tau)\rangle$, where

$$\theta_{\Delta t}(\tau) = \frac{1}{\Delta t}[k(x - \tau, t + \Delta t) - k(x - \tau, t)] - D_t k(x - \tau, t), \qquad \Delta t \neq 0.$$

We will have established (5.3) when we show that, as $\Delta t \to 0$, $\theta_{\Delta t}(\tau) \to 0$ in $\mathfrak{W}_{a,b}$ for every choice of a and b $(a < b)$. By using the identity

$$(5.4) \qquad D_\tau^2 k(x - \tau, t) = D_t k(x - \tau, t),$$

one can show that

$$D_\tau^n \theta_{\Delta t}(\tau) = \frac{1}{\Delta t}\int_0^{\Delta t} dy \int_0^y D_\tau^{n+4} k(x - \tau, t + z)\, dz,$$

from which the desired result follows after some computation.

Equations (5.2), (5.3), and (5.4) now imply that $D_x^2 u(x, t) = D_t u(x, t)$ for $-\infty < x < \infty, 0 < t < 1$.

We finally show that, for any $\phi \in \mathfrak{D}$, $\langle u(x, t), \phi(x)\rangle \to \langle f, \phi\rangle$ as $t \to 0+$. Since $\phi \in \mathfrak{D}$ and since $k(x - \tau, t)$ satisfies the hypothesis on $\theta(\tau, x)$ in Lemma 2.1 for every a and b, we may invoke this lemma to write

$$\langle u(x, t), \phi(x)\rangle = \langle \phi(x), \langle f(\tau), k(x - \tau, t)\rangle\rangle = \langle f(\tau), \langle \phi(x), k(x - \tau, t)\rangle\rangle.$$

We have already demonstrated in the last part of the proof of Theorem 4.1 that, as $t \to 0+$, $\langle \phi(x), k(x - \tau, t)\rangle \to \phi(\tau)$ in $\mathfrak{W}_{a,b}$ for every a and b. This completes the proof.

In the foregoing analysis the time variable t was restricted to the interval $0 < t < 1$ because we had set $c = 1$ in the expression $\exp(-cs^2)$; see (1.4). Our results can be extended to any time interval $0 < t < c$ by replacing t by t/c and then making appropriate changes of variables in the above formulas and definitions.

6. Generalized convolution transformations with class E kernels. A kernel G is said to be in class E if it is defined by (1.3) and (1.4); here, c need not be zero. When the convolution transformation (1.2) possesses such a kernel (with $c = 1$), we can extend it to certain generalized functions by combining the foregoing results with those of [5]. In particular, let $\mathfrak{L}'_{\gamma,\delta}$ be the space of generalized functions defined in [5, §3], and let H be a kernel of the form (1.3) wherein $c = 0$. Also, set

$$\alpha_1 = \max_{a_\nu<0} (a_\nu, -\infty), \qquad \alpha_2 = \min_{a_\nu>0} (a_\nu, +\infty). \tag{6.1}$$

If $f \in \mathfrak{L}'_{\gamma,\delta}$ for some γ and δ such that $\gamma < \alpha_2$ and $\delta > \alpha_1$, then the convolution transform

$$\tilde{f}(\tau) = \langle f(y), H(\tau - y)\rangle, \qquad -\infty < \tau < \infty, \tag{6.2}$$

is a member of the testing function space $\mathfrak{L}_{\alpha,\beta}$ defined in [5, §3], where $\alpha < \min(-\alpha_1, -\gamma)$ and $\beta > \max(-\alpha_2, -\delta)$ [5, Theorem 4.1]. By note 9 of §2, and by the definition of $\mathfrak{L}_{\alpha,\beta}$, $\tilde{f}(\tau)$ generates a member of $\mathfrak{W}'_{a,b}$ for every a and b ($a < b$). Thus, we may write

$$F(s) = \langle \tilde{f}(\tau), k(s - \tau, 1)\rangle, \qquad -\infty < \operatorname{Re} s < \infty. \tag{6.3}$$

Consequently, when G is a class E kernel with $c = 1$, we can define the convolution transform of $f \in \mathfrak{L}'_{\gamma,\delta}$ by combining (6.2) and (6.3):

$$F(s) = \langle\langle f(y), H(\tau - y)\rangle, k(s - \tau, 1)\rangle, \qquad -\infty < \operatorname{Re} s < \infty, \tag{6.4}$$

where

$$H(\tau) = \frac{1}{2\pi i}\int_{-i\infty}^{i\infty} \frac{e^{-z^2} e^{z\tau}}{E(z)}\, dz. \tag{6.5}$$

An inversion formula for (6.4) is given by Theorem 6.1.

THEOREM 6.1. *Let $E(z)$ be defined by (1.4) with $c = 1$, let $H(\tau)$ be defined by (6.5), and let α_1 and α_2 be defined by (6.1). If $f \in \mathfrak{L}'_{\gamma,\delta}$, where $\gamma < \alpha_2$ and $\delta > \alpha_1$, and if $F(s)$ is defined by (6.4), then, in the sense of weak convergence in $\mathfrak{D}'$,*

$$\lim_{n\to\infty} \lim_{t\to 1-} P_n(D_\tau) \int_{-\infty}^{\infty} k(\omega + i\tau - i\sigma, t)F(\sigma + i\omega)\, d\omega = f(\tau),$$

where

$$P_n(D) = e^{(b-b_n)D} \prod_{\nu=1}^{n} \left(1 - \frac{D}{a_\nu}\right) e^{D/a_\nu},$$

and $\{b_n\}_{n=1}^{\infty}$ is any sequence of real numbers tending to zero.

This follows directly from Theorem 4.1 and [5, Theorem 5.3].

Acknowledgment. The author is indebted to T. P. G. Liverman for several criticisms and suggestions which improved the original version of this work.

REFERENCES

[1] I. I. HIRSCHMAN AND D. V. WIDDER, *The Convolution Transform*, Princeton University Press, Princeton, 1955.

[2] P. G. ROONEY, *A generalization of an inversion formula for the Gauss transformation*, Canad. Math. Bull., 6 (1963), pp. 45–53.

[3] E. HILLE AND R. S. PHILLIPS, *Functional Analysis and Semi-groups*, Colloquium Publications, vol. XXXI, rev. ed., American Mathematical Society, Providence, 1957.

[4] A. GONZALEZ DOMINGUEZ, *A contribution to the theory of Hille functions*, Ciencia y Técnica, 42 (1941), pp. 283–329.

[5] A. H. ZEMANIAN, *A generalized convolution transformation*, this Journal, 15 (1967), pp. 324–346.

[6] L. SCHWARTZ, *Théorie des distributions*, vols. I and II, Hermann, Paris, 1957 and 1959.

[7] A. FRIEDMAN, *Generalized Functions and Partial Differential Equations*, Prentice-Hall, Englewood Cliffs, New Jersey, 1963.

[8] A. H. ZEMANIAN, *Distribution Theory and Transform Analysis*, McGraw-Hill, New York, 1965.

A SOLUTION OF A DIVISION PROBLEM ARISING FROM BESSEL-TYPE DIFFERENTIAL EQUATIONS*

A. H. ZEMANIAN†

1. In a recent work [1] the Hankel transformation was extended to certain generalized functions on the half-line $0 < x < \infty$ having no restriction on their rate of growth as $x \to \infty$. The present work is devoted to an application of this generalized Hankel transformation; in particular, it provides a fairly straightforward method for finding the solution u to the differential equation

$$P(S_\mu) = g, \tag{1}$$

where g is any generalized function whose Hankel transform of order μ exists in the sense of [1], P is a polynomial, μ is a real parameter $\geqq -\frac{1}{2}$, and

$$S_\mu = x^{-\mu-1/2}Dx^{2\mu+1}Dx^{-\mu-1/2}. \tag{2}$$

Letting η denote a complex variable and applying the μth order Hankel transformation to (1), we get

$$P(-\eta^2)U(\eta) = G(\eta), \tag{3}$$

where $G(\eta)$ is now a generalized function on a certain space of analytic functions. Thus, our problem becomes one of dividing $G(\eta)$ by $P(-\eta^2)$ to obtain a generalized μth-order Hankel transform $U(\eta)$.

In this work our symbols have precisely the same meaning as in [1]; we do not repeat their definitions here. On the other hand, we alter our terminology in just one way: in [1] we referred to the members of $\mathcal{B}_\mu'$ as "distributions" even though, as was pointed out there, $\mathcal{B}_\mu'$ cannot be identified in a one-to-one fashion with a subset of $\mathfrak{D}_I'$. Here, however, we call the members of $\mathcal{B}_\mu'$ *generalized functions* since this conforms to the more common practice of restricting the name "distribution" to the members of $\mathfrak{D}_I'$.

2. Assume that g is a generalized function in the space $\mathcal{B}_\mu'$ ($\mu \geqq -\frac{1}{2}$). We shall first solve (1) under the assumption that the polynomial P does not have a root at the origin. The case where $P(0) = 0$ requires a special treatment and is considered subsequently. An application of the generalized

* Received by the editors August 12, 1966. Contributed at the Symposium on "The Applications of Generalized Functions" sponsored by the Air Force Office of Scientific Research at the 1966 Fall Meeting of Society for Industrial and Applied Mathematics held at the State University of New York at Stony Brook, September 12–14, 1966.

† Department of Applied Analysis, State University of New York at Stony Brook, Stony Brook, Long Island, New York 11790. This work was supported by the Air Force Cambridge Research Laboratories, Bedford, Massachusetts, under Contract AF19 (628)-2981.

μth-order Hankel transformation to (1) yields (3) where $G(\eta) \in \mathcal{Y}_\mu'$. We can find a $U(\eta) \in \mathcal{Y}_\mu'$ which satisfies (3) by modifying a known procedure [2, pp. 123–126].

Let the roots of $P(-\eta^2)$ in the η-plane be denoted by γ_n, $n = \pm 1, \cdots, \pm q$, $\gamma_n = \gamma_{-n} \neq 0$, $k_n = k_{-n}$. Thus,

$$P(-\eta^2) = \alpha_m(\eta^2 - \gamma_1^2)^{k_1} \cdots (\eta^2 - \gamma_q^2)^{k_q},$$

where $\alpha_m \neq 0$ is a constant. It follows directly from the definition of the space $\mathcal{Y}_\mu$ that $P(-\eta^2)$ is a multiplier in $\mathcal{Y}_\mu$. Consequently, in view of (3) we may write, for any $\Phi(\eta) \in \mathcal{Y}_\mu$,

$$\langle U(\eta), P(-\eta^2)\Phi(\eta)\rangle = \langle G(\eta), \Phi(\eta)\rangle.$$

This determines U on every $\mathrm{X} \in \mathcal{Y}_\mu$ of the form $\mathrm{X}(\eta) = P(-\eta^2)\Phi(\eta)$, where $\Phi \in \mathcal{Y}_\mu$; namely,

$$\langle U, \mathrm{X}\rangle = \left\langle G(\eta), \frac{\mathrm{X}(\eta)}{P(-\eta^2)}\right\rangle. \tag{4}$$

According to [1, Lemma 26], X has the stated form if and only if

$$\mathrm{X}^{(\nu)}(\pm\gamma_n) = 0, \qquad \nu = 0, 1, \cdots, k_n - 1, \quad n = 1, 2, \cdots, q. \tag{5}$$

Moreover, by [1, Lemma 27], for any choice of the entire functions Λ_n, $n = \pm 1, \cdots, \pm q$, which satisfy the hypothesis of that lemma, we can uniquely decompose any $\Psi \in \mathcal{Y}_\mu$ into the form [1, (16)]. By applying $U(\eta)$ to this decomposition of Ψ, making use of (4), and arbitrarily assigning numerical values to

$$\langle U(\eta), \eta^{u+1/2}[\Lambda_n(\eta)(\eta - \gamma_n)^\nu + (-1)^\nu\Lambda_{-n}(\eta)(\eta - \gamma_{-n})^\nu]\rangle,$$

we obtain

$$\langle U, \Psi\rangle = \left\langle G(\eta), \frac{\mathrm{X}(\eta)}{P(-\eta^2)}\right\rangle + \sum_{n=1}^{q}\sum_{\nu=0}^{k_n-1} b_{n,\nu}\Psi^{(\nu)}(\gamma_n), \tag{6}$$

where the constants $b_{n,\nu}$ are arbitrary. The double summation here is the value assigned to $\Psi(\eta)$ by the sum of delta functionals

$$\sum_{n=1}^{q}\sum_{\nu=0}^{k_n-1} (-1)^\nu b_{n,\nu}\delta^{(\nu)}(\eta - \gamma_n),$$

which is clearly a member of $\mathcal{Y}_\mu'$. Thus, (6) determines U as a functional on $\mathcal{Y}_\mu$. We can prove that U is a member of $\mathcal{Y}_\mu'$ by showing that the first term on the right-hand side of (6) defines a continuous linear functional on $\mathcal{Y}_\mu$.

Linearity follows directly from the fact that the decomposition [1, (16)] is a linear one. To prove continuity, let $\{\Psi_m\}_{m=1}^\infty$ converge in $\mathcal{Y}_\mu$ to zero. Since $G \in \mathcal{Y}_\mu'$, we need prove merely that, for fixed choices of the Λ_n in the decomposition of the Ψ_m in accordance with [1, (16)], the sequence $\{X_m(\eta)/P(-\eta^2)\}_{m=1}^{\infty}$ also converges in $\mathcal{Y}_\mu$ to zero, because this will then imply the convergence of $\langle U, \Psi_m\rangle$ to zero.

If $\Psi_m \to 0$ in $\mathcal{Y}_\mu$ as $m \to \infty$, then $\eta^{-\mu-1/2}\Psi_m(\eta) \to 0$ uniformly on every bounded domain of the η-plane. Hence, from the decomposition [1, (16)] it follows that $\{\eta^{-\mu-1/2}X_m(\eta)/P(-\eta^2)\}_{m=1}^{\infty}$ converges to zero uniformly on any compact domain of the η-plane that contains no zeros of $P(-\eta^2)$.

Next, let C_0 denote a circle with center at γ_1 and radius $3R$ which does not contain the origin or any other zero of $P(-\eta^2)$ within its interior. Let C be a concentric circle of radius $2R$. Consider

$$\begin{aligned}(7)\qquad &\frac{\eta^{-\mu-1/2}X_m(\eta)}{P(-\eta^2)}\\ &= \frac{1}{P(-\eta^2)}\left\{\eta^{-\mu-1/2}\Psi_m(\eta) - \sum_{\nu=0}^{k_1-1}\frac{1}{\nu!}\hat{\Psi}^{(\nu)}(\gamma_1)\Lambda_1(\eta)(\eta-\gamma_1)^{\nu}\right\} + H_m(\eta)\end{aligned}$$

where $H_m(\eta)$ consists of all the other terms in the decomposition of $\Psi_m(\eta)$ multiplied by $-\eta^{-\mu-1/2}/P(-\eta^2)$. Since every $\Lambda_n(\eta)$ $(n \neq 1)$ has a zero at $\eta = \gamma_1$ whose multiplicity is at least k_1, $\Lambda_n(\eta)/P(-\eta^2)$ is analytic within and on C. Moreover, $\Psi_m^{(\nu)}(\gamma_n) \to 0$ as $m \to \infty$, and consequently $H_m(\eta) \to 0$ uniformly within and on C.

Furthermore, set $P(-\eta^2) = (\eta - \gamma_1)^{k_1}Q(\eta)$. Hence $Q(\eta)$ is a polynomial having no zeros within C_0. By Taylor's formula with remainder [3, p. 130], we can rewrite the first term containing the braces on the right-hand side of (7) in the following way, where we now restrict η by $|\eta - \gamma_1| < R$:

$$\begin{aligned}(8)\qquad &\frac{1}{Q(\eta)}\sum_{\nu=0}^{k_1-1}\frac{1}{\nu!}\hat{\Psi}^{(\nu)}(\gamma_1)[1-\Lambda_1(\eta)](\eta-\gamma_1)^{\nu-k_1}\\ &\qquad + \frac{1}{2\pi i Q(\eta)}\int_C \frac{\zeta^{-\mu-1/2}\Psi_m(\zeta)}{(\zeta-\eta)(\zeta-\gamma_1)^{k_1}}\,d\zeta.\end{aligned}$$

But, $1 - \Lambda_1(\eta)$ also has a zero at γ_1 whose multiplicity is at least k_1. Therefore, the $[1 - \Lambda_1(\eta)](\eta - \gamma_1)^{\nu-k_1}$, $\nu = 0, \cdots, k_1 - 1$, are bounded for $|\eta - \gamma_1| < R$, which implies that the first term in (8) converges uniformly to zero for $|\eta - \gamma_1| < R$ in view of the fact that $\hat{\Psi}_m^{(\nu)}(\gamma_1) \to 0$ as $m \to \infty$. On the other hand, for $|\eta - \gamma_1| < R$, we have $|\zeta - \eta| > R$ so that the magnitude of the second term in (8) is bounded by

$$\frac{2}{|Q(\eta)|\,(2R)^{k_1}}\sup_{\zeta\in C}|\zeta^{-\mu-1/2}\Psi_m(\zeta)|,$$

which tends to zero as $m \to \infty$ uniformly for $|\eta - \gamma_1| < R$.

Thus, we have proved that (7) converges uniformly to zero for $|\eta - \gamma_1| < R$. The same is true in neighborhoods of every zero of $P(-\eta^2)$

since we can choose γ_1 to be any such zero. By what was shown previously, therefore, (7) converges uniformly to zero on every bounded domain of the η-plane.

Since $\{\Psi_m\}$ converges to zero in $\mathcal{Y}_\mu$, it converges to zero in $\mathcal{Y}_{\mu,b}$ for some $b > 0$. All that remains to be proven is that

$$(9) \qquad \left| e^{-b|\omega|} \eta^{2k-\mu-1/2} \frac{X_m(\eta)}{P(\eta^2)} \right| < C_k$$

for all η, where the constants C_k do not depend on m. Our arguments have already shown that (9) holds true on every bounded domain of the η-plane. So, let Γ be an open set in the η-plane which contains all the zeros of $P(-\eta^2)$, and let Ξ be the complement of Γ. Then, $1/P(-\eta^2)$ is bounded on Ξ. Also, from the definition of convergence in $\mathcal{Y}_{\mu,b}$,

$$| e^{-b|\omega|} \eta^{2k-\mu-1/2} \Psi_m(\eta) | < B_k$$

for all m and all η. Moreover,

$$e^{-b|\omega|} \eta^{2k} [\Lambda_n(\eta)(\eta - \gamma_n)^\nu + (-1)^\nu \Lambda_{-n}(\eta)(\eta - \gamma_n)^\nu]$$

is also bounded on Ξ by Condition 1 of the hypothesis of [1, Lemma 27]. In view of (7), this shows that (9) is indeed satisfied. Thus we have proven that as $m \to \infty$, $X_m(\eta)/P(-\eta^2) \to 0$ in $\mathcal{Y}_\mu$ whenever $\Psi_m(\eta) \to 0$ in $\mathcal{Y}_\mu$. This completes the proof of the fact that U, as defined by (6), is a member of $\mathcal{Y}_\mu'$.

$U(\eta)$ is the solution to the division problem (3) where $P(0) \neq 0$. By means of the inverse Hankel transformation we now find that solutions u to the differential equation (1) with $g \in \mathcal{B}_\mu'$ and $P(0) \neq 0$ are specified on any $\psi \in \mathcal{B}_\mu$ by $\langle u, \psi \rangle = \langle U, \Psi \rangle$, where $\Psi = \mathfrak{H}_\mu \psi$ and $\langle U, \Psi \rangle$ is (6). Actually, these are the only solutions because all solutions to the homogeneous equation $P(M_\mu N_\mu)u = 0$ take the form

$$\sum_{n=1}^{q} \sum_{\nu=0}^{k_n - 1} b_{n,\nu} D_\eta^\nu [\sqrt{x\eta}\, J_\mu(x\eta)] |_{\eta=\gamma_n},$$

which accounts for the double summation in (6); this was demonstrated in [1, §§6 and 7].

3. We now remove the restriction that our polynomial of differentiation have no root at the origin. In particular, we shall seek a solution to

$$(10) \qquad S_\mu^{k_0} P(S_\mu) u = g,$$

where $g \in \mathcal{B}_\mu'$ ($\mu \geqq -\frac{1}{2}$), $P(-\eta^2)$ is a polynomial with $P(0) \neq 0$, k_0 is a positive integer, and the unknown u is required to be in $\mathcal{B}_\mu'$ also. We shall make one further assumption: g is such that $G = \mathfrak{H}_\mu g$ is a member of $\mathcal{Y}'_{\mu-2k_0}$. More precisely, we assume that there exists a $\tilde{G} \in \mathcal{Y}'_{\mu-2k_0}$ whose restriction to $\mathcal{Y}_\mu$ is equal to $G = \mathfrak{H}_\mu g$. If one such $\tilde{G}$ exists, there will be an infinity of such $\tilde{G}$, as we shall see.

An application of the generalized Hankel transformation of order μ to (10) yields

$$(11) \qquad (-\eta^2)^{k_0} P(-\eta^2) U(\eta) = G(\eta).$$

By [1, Lemma 21], the generalized function

$$(12) \qquad P(-\eta^2) U(\eta) = (-\eta^2)^{-k_0} \tilde{G}(\eta)$$

is a member of $\mathcal{Y}_\mu'$; also, it satisfies (11) in the sense of equality in $\mathcal{Y}_\mu'$. Moreover, we may add to (12) the general solution $H(\eta)$ of the homogeneous equation $(-\eta^2)^{k_0} H(\eta) = 0$:

$$(13) \qquad H(\eta) = \sum_{\nu=0}^{k_0-1} \alpha_\nu F^{(2\nu)}(\eta),$$

where the α_ν are arbitrary constants and $F^{(2\nu)}(\eta) \in \mathcal{Y}_\mu'$ is defined by

$$\langle F^{(2\nu)}(\eta), \Phi(\eta) \rangle = \lim_{\eta \to 0} D^{2\nu}[\eta^{-\mu-1/2} \Phi(\eta)], \qquad \Psi \in Y_\mu .$$

(Note that the odd-order derivatives of $\eta^{-\mu-1/2}\Phi(\eta)$ vanish at $\eta = 0$ since $\eta^{-\mu-1/2}\Phi(\eta)$ is an even entire function of η defined by continuity at $\eta = 0$.) There are no solutions to $(-\eta^2)^{k_0} H(\eta) = 0$ other than (13). The proof of this is almost identical to the proof of the corresponding assertion for the space Z' (see [4, §7.10] and [1, §7]). Thus, the most general form for $P(-\eta^2) U(\eta)$ as a solution to (11) is

$$(14) \qquad P(-\eta^2) U(\eta) = (-\eta^2)^{-k_0} \tilde{G}(\eta) + \sum_{\nu=0}^{k_0-1} \alpha_\nu F^{(2\nu)}(\eta).$$

Now, we can solve (14) for $U(\eta)$ by the method of §2. Thus, decomposing any $\Psi \in Y_\mu$ according to [1, (16)], having first fixed the Λ_n, from (6) we obtain

$$(15) \qquad \begin{aligned} \langle U(\eta), \Psi(\eta) \rangle = \left\langle (-\eta^2)^{-k_0} \tilde{G}(\eta) + \sum_{\nu=0}^{k_0-1} \alpha_\nu F^{(2\nu)}(\eta), \frac{\mathrm{X}(\eta)}{P(-\eta^2)} \right\rangle \\ + \sum_{n=1}^{q} \sum_{\nu=0}^{k_n-1} b_{n,\nu} \Psi^{(\nu)}(\gamma_n). \end{aligned}$$

By the definition of our generalized Hankel transformation, the solutions and only solutions in $\mathcal{B}_\mu'$ of (10) are those members u of $\mathcal{B}_\mu'$ that assign to any $\psi \in \mathcal{B}_\mu$ the values (15), wherein $\Psi = \mathfrak{H}_\mu \psi$ and α_ν and $b_{n,\nu}$ are arbitrarily chosen constants.

REFERENCES

[1] A. H. ZEMANIAN, *The Hankel transformation of certain distributions of rapid growth*, this Journal, 14 (1966), pp. 678–690.
[2] L. SCHWARTZ, *Théorie des distributions*, vol. I, Hermann, Paris, 1957.
[3] R. V. CHURCHILL, *Complex Variables and Applications*, 2nd ed., McGraw-Hill, New York, 1960.
[4] A. H. ZEMANIAN, *Distribution Theory and Transform Analysis*, McGraw-Hill, New York, 1965.